Fiber-Reinforced Plastics

Edited by Martin Alberto Masuelli

Published in London, United Kingdom

IntechOpen

Supporting open minds since 2005

Fiber-Reinforced Plastics
http://dx.doi.org/10.5772/intechopen.95632
Edited by Martin Alberto Masuelli

Contributors
Svetlana Risteska, Junhua Wei, Archana Nigrawal, Arun Kumar Sharma, Fozia Zia Haque, Bikash Chandra Chakraborty, Allana Azevedo do Nascimento, Carlos José Salvador Tomassini, Irene Samy, Liqaa Hamid, Ramratan Guru, Rohit Kumar, Anupam Kumar, Nicholas Lambrache, Ora Renagi, Lidia Olaru, Brian N'Drelan, Paul Nestor Djomou Djonga, Ahmat Tom, Hambate Gomdje Valery, Georges Elambo Nkeng

First published in London, United Kingdom, 2022 by IntechOpen
IntechOpen is the global imprint of INTECHOPEN LIMITED, registered in England and Wales, registration number: 11086078, 5 Princes Gate Court, London, SW7 2QJ, United Kingdom
Printed in Croatia

British Library Cataloguing-in-Publication Data
A catalogue record for this book is available from the British Library

Additional hard and PDF copies can be obtained from orders@intechopen.com

Fiber-Reinforced Plastics
Edited by Martin Alberto Masuelli
p. cm.
Print ISBN 978-1-80355-075-6
Online ISBN 978-1-80355-076-3
eBook (PDF) ISBN 978-1-80355-077-0

Meet the editor

Martin A. Masuelli is an invited adjunct professor at the Instituto de Física Aplicada, National Council for Scientific and Technical Research (CONICET), National University of San Luis (UNSL), Argentina, and an associate professor at UNSL, where he obtained his master's degree and Ph.D. in Membrane Technology. He is also the director of the Physics Chemistry Service Laboratory, UNSL. Dr. Masuelli is an expert in polysaccharides and physics chemistry of macromolecules. He has authored or co-authored more than thirty-two peer-reviewed international publications, eight book chapters, and seventy communications in international congresses. He has also edited seven books. He is a member of the Sociedad Argentina de Ciencia y Tecnología Ambiental, Asociación Argentina de Fisicoquímica y Química Inorgánica, and Asociación Argentina de Tecnólogos de Alimentos. He is the editor in chief and founder of the *Journal of Polymer and Biopolymers Physics Chemistry* and an editorial board member for numerous other journals. Dr. Masuelli's research interests include hydropolymers and biopolymers (separative processes, purification processes, and characterization), physiochemistry of macromolecules, membrane technology and design (NF-UF-MF), and separative processes.

Contents

Preface

Fiberglass reinforced plastic (FRP) is a composite material made up of a matrix of plastic or polymeric resin reinforced with glass fibers (GRP), carbon fibers (CRP), natural fibers (NRP), etc. The existence of these modern materials dates to the petroleum age more than 100 years ago.

FRP is a light, resistant, and easy to mold material, which is why it is widely used in the manufacture of parts with soft and complex shapes. This material has a wide variety of applications and functionalities in the nautical, military, commercial, and industrial industries.

Fiberglass reinforcement provides the composite with mechanical strength, dimensional stability, and heat resistance. Polyester plastic resin provides dielectric chemical resistance and outdoor behavior.

For natural fiber-reinforced plastics (NRP), thermoplastic or thermosetting polymers are used as the matrix. Thermoplastics are usually PE, PP, PS, PVC, or PLA, and acrylate, epoxy, and phenolic resins are generally used as thermosetting. Fibers include flax, hemp, jute, kenaf, sisal, abaca, cellulose, and lignocellulose of various origins, and cotton. Other widely used materials include wood fiber and wood flour or rice husk. In many cases, specific coupling agents are used to ensure the optimal union of matrix and fibers.

Carbon fiber (CFR) is a material made up of fibers 5–10 microns in diameter, composed mainly of carbon atoms. The carbon atoms are bonded together in crystals that are aligned parallel to the long axis of the fiber. Crystal alignment gives the fiber high strength based on bulk (makes it strong for its size). Several thousand carbon fibers are twisted together to form a thread, which can be used on its own or woven into fabric. The properties of carbon fibers, such as high flexibility, high strength, low weight, high temperature tolerance, and low thermal expansion, make them very popular in aerospace, civil engineering, military, motor sports, and many other applications. However, they are relatively expensive compared to similar fibers, such as glass fibers or plastic fibers, which greatly limits their use.

This book presents a comprehensive overview of the different types of reinforced plastic fibers for engineers, researchers, academics, students, and industry professionals.

This book includes the following sections and chapters:

Section 1: "Polymers, Composites and Thermoplastics"

Chapter 1: "Self-Healing Polymers and Composite Materials" by Allana Azevedo do Nascimento.

Chapter 2: "Design, Simulation, and Analysis of the Extrusion Process of a PVC Thermoplastic Profile to Optimize the Design of the Die and the Machine Parameters" by Carlos José Salvador Tomassini.

Section 2: "Carbon Fiber Reinforced"

Chapter 3: "Unidirectional Carbon Fiber Reinforced Thermoplastic Tape in Automated Tape Placement Process" by Svetlana Risteska.

Chapter 4: "Mechanically Improved and Multifunctional CFRP Enabled by Resins with High Concentrations Epoxy-Functionalized Fluorographene Fillers" by Junhua Wei

Section 3: "Natural Fiber Reinforced"

Chapter 5: "Composite Materials with Natural Fibers" by Nicholas Lambrache, Ora Renagi, Lidia Olaru, and Brian N'Drelan.

Chapter 6: "Sisal Fibre Based Polymeric Composites" by Archana Nigrawal, Arun Kumar Sharma and Fozia Zia Haque

Chapter 7: "Functional Application for the Corn Leaf Fibre to Make Reinforced Polymer Composites Sheet" by Ramratan Guru, Anupam Kumar and Rohit Kumar.

Section 4: "FRP Application"

Chapter 8: "Fabricating Natural Biocomposites for Food Packaging" by Liqaa Hamid and Irene Samy.

Chapter 9: "FRP for Marine Application" by Bikash Chandra Chakraborty.

Chapter 10: "Mechanical Properties and Chemical Stability of Bathroom Wall Composites Manufactured from Recycle Polyethylene Terephthalate (PET) Mixed with Cocoa Hull Powder" by Paul Nestor Djomou Djonga, Ahmat Tom, Hambate Gomdje Valery and Georges Elambo Nkeng.

Martin Alberto Masuelli
Associate Professor Sup,
Area of Physical Chemistry,
Department of Chemistry,
Faculty of Chemistry,
Biochemistry and Pharmacy,
National University of San Luis. Adj.
Researcher-Institute of Applied Physics-CONICET.
Director of the Physical Chemistry Research
and Services Laboratory (LISEQF-UNSL),
San Luis, Argentina

Section 1

Polymers, Composites and Thermoplastics

Chapter 1

Self-Healing Polymers and Composite Materials

Allana Azevedo do Nascimento

Abstract

In order to overcome the challenges and limitations related to conventional maintenance and repair methods in structural composites during service, the concept of self-healing for polymeric materials has been developed in the last decades. Inspired by biological systems, ideal self-healing materials must be able to repair damages continuously during the service life of the component, recovering its performance. Several techniques have been proposed in the last years to manufacture self-healing polymers and fiber-reinforced composites to provide healing of microcracks in the composite structure without or with less intervention, extending service life and safety of the components and reducing maintenance time and cost. This book chapter proposes an overview of the most promising self-healing approaches for thermoset and polymer matrix composites developed in recent year.

Keywords: advanced materials, thermoset polymers, polymer matrix composites, self-healing, mendable resin

1. Introduction

Polymers and polymer matrix composites (PMCs) have been extensively used for a variety of structural applications due their combination of high mechanical properties and light weight savings as major benefit. During service, the operating environment often exposes these materials to severe variations in mechanical loading and environmental conditions. Such conditions induce damages and can cause the formation of microcracks. These matrix damage can lead to the coalesce and growth of larger cracks, which compromise mechanical performance and integrity of the material inside its structure, where detection and external intervention are difficult and/or impracticable. Non-destructive testing methods are generally used for inspection of components during operation. However, internal microcracks can be difficult to repair [1, 2]. To overcome the challenges related to the repair of these internal damages in thermoset matrix and FRP, self-healing materials have been proposed in the decades as an alternative to conventional repair methods.

Ideal self-healing materials are designed to present the ability repair themselves, using resources inherently available to them. The recovery of performance can be full or partial and occurs during the service life of the component with minimal or no external intervention. These materials can provide the healing of invisible microcracks in the composite structure and can be an alternative to non-effective conventional maintenance methods to extend service life and safety of the components in addition to reducing maintenance time and cost [1, 3, 4].

Inspired by biological systems, the self-healing of synthetic materials occurs in three steps. Initially, the healing mechanism is triggered by the damage. In the second step, the healing agent, that is, the material responsible to repair the damage, is transported to the damaged area. Lastly, the repair process takes place, recovering the crack. The time and efficiency of the recovery depend on the self-healing mechanism used [5, 6].

Different types of self-healing mechanisms are described in literature for all types of materials. Polymeric materials are reported as promising for self-healing, and the self-healing ability can be introduced by chemical modification or exposure to physical or thermal conditions. The glass transition temperature (Tg), the chain mobility, and the chemical structure are important parameters to be considered during the design of the self-healing system and can influence the success of the healing ability [7].

In this scenario, self-healing polymeric materials represent a new class of materials and have attracted great attention in the past decades. Studies have been developed to better understanding of the self-healing ability, the mechanisms, and how to overcome the challenges of processing and the limitations of each system. In the last decades, a significant increase in the amount of research about self-healing polymers is reported. **Figure 1** shows the growth of the number of papers published on the topic of self-healing polymeric materials since 2000.

Several mechanisms are employed to obtain self-healing polymeric materials and each mechanism presents advantages, limitations, and challenges in formulation. In this review article, principles and developments in self-healing thermoset matrix and fiber reinforced polymers were presented.

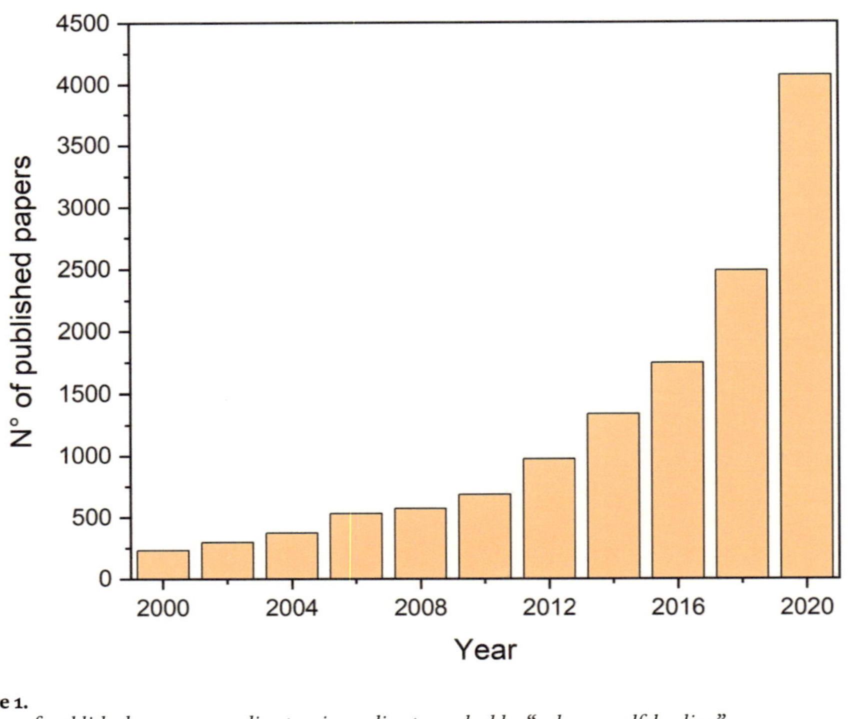

Figure 1.
Number of published papers according to science direct searched by "polymers self-healing".

2. Self-healing mechanisms

Self-healing polymeric materials are capable of recovering their properties due to the repair of microcracks. Different mechanisms can be used to confer the inherent ability to self-heal to polymers. For self-healing materials, the recovery of properties can occur naturally (autonomic self-healing) or be activated after the application of an external stimulus such as heat or radiation (non-autonomic self-healing) [6, 8]. Different mechanisms of self-healing for polymer composites have been reported over the last decade. Each mechanism has limitations and aspects to be improved. Generally, self-healing mechanisms can be classified into two main groups: intrinsic and extrinsic self-healing mechanisms [9].

For extrinsic self-healing mechanisms, the healing agent is inserted into containers and incorporated into the matrix. Microcapsules and vascular network are the main extrinsic self-healing mechanisms. For encapsulated healing agents, during crack formation, the capsules are broken, releasing the healing agent within containers to fill the cracks. The polymerization of the healing agent will occur in these regions and heal the crack plane. Self-healing of polymer composites using the vascular network mechanism is developed by incorporating pipelines filled with a healing agent within a polymer composite matrix. When damage is triggered, the healing agent flows into the damage region by capillarity and polymerizes, thereby completing the process. These extrinsic mechanisms provide recovery of up to 100% of the initial fracture resistance of epoxy resins. However, despite the high recovery levels, the difficulty of processing and the inability to multiply cycles of healing in the same area are limiting factors for the application of this mechanism on a large scale [6, 8].

Intrinsic mechanisms are defined by the inherent ability to healing of the matrix, after a damage, in the presence of external stimulus such UV light, heat, or chemicals. Thermoset polymer matrices with intrinsic healing capability, also known as mendable resins, have the advantage of multiple healing cycles as long as damage to the reinforcement fibers is not excessive or highly localized. Different mechanisms have been proposed in the last decade for the production of self-healing polymer matrices [6, 8]. **Figure 2** summarizes the types of self-healing polymeric materials.

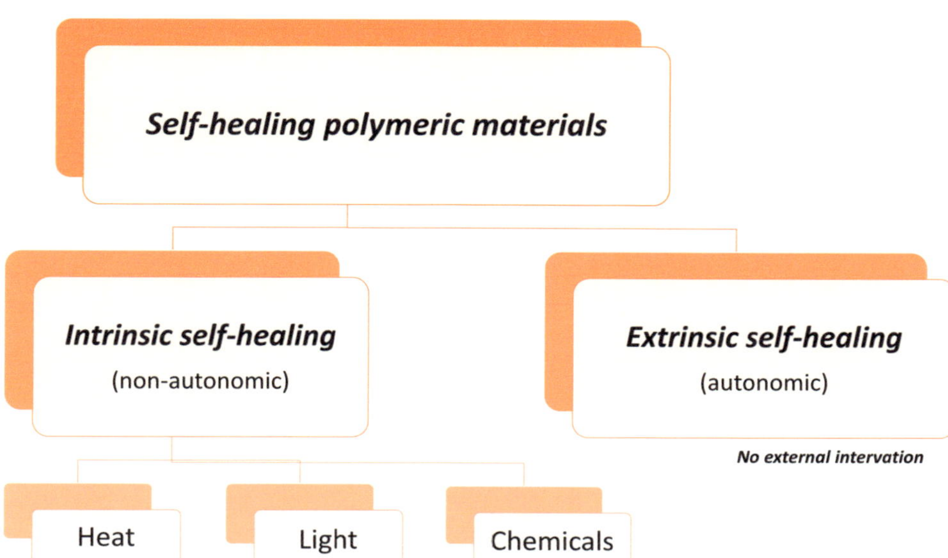

Figure 2.
Types of self-healing polymeric materials.

The healing of polymeric materials may be related to the recovery of several properties of the material, including fracture toughness, tensile strength, barrier properties and even molecular weight [1]. Then, the extent of healing can be difficult to compare, due to the range of properties that are healed in these materials. However, the majority of studies have used a basic method for describing the extent of healing in polymeric systems for a range of properties. The healing efficiency (η) is determined by the ratio of the value of the property tested after and before the healing (Eq. (1)) 5. This equation has been used as the basis for a non-property-specific method of comparing the healing efficiency of different self-healing polymeric systems [1, 3, 4].

$$Healing\ efficiency\left(\eta_p\right)=\frac{Property\ value\ healed}{Property\ value\ initial}\times 100 \quad (1)$$

For self-healing efficiency assessment, the protocol reported on the literature involve basically three steps: (i) submit the test specimens to initial tests to promote controlled damage, (ii) healing (that can be triggered autonomically or non-autonomically), and (iii) submit the healed specimens to tests again, in order to evaluate the new properties. Usually, the first tests are carried out until a certain level of load provides damage but not the failure of the samples, while the second test is carried out until total failure. The ratio between the properties measured before and after damage is then used to quantify the healing efficiency. Nonetheless, it is not unusual to find self-healing reports that involve the visual observation of microcracks healing, without quantification of healing efficiency.

Self-healing efficiency greater than 100% have been reported for thermoset matrix in several self-healing studies. However, results of unmodified control samples are often not shown as the basis for comparison and the healing protocols are not standardized. In addition, the properties of the modified materials should be reported. Hence, although the healing efficiency equation has been extensively used for self-healing studies, a need for standardized methods for self-healing measurements still exists. In addition, when results of healing efficiencies obtained through different methods were compared it was found that healing protocols can be highly inaccurate and therefore not recommended [10, 11].

2.1 Extrinsic self-healing

Considered autonomic self-healing, extrinsic self-healing mechanisms are usually classified according to the type of the storage vessels used, although the self-healing concept is similar. Capsuled-based self-healing and vascular network self-healing are the two types of extrinsic self-healing mechanisms [6, 8].

2.1.1 Capsule-based self-healing

For capsule-based self-healing polymers, the healing agent is sequestered into microcapsules until the damage. The cracks formed are responsible for the rupture of the capsules, releasing the healing agent and healing the crack plane. This mechanism was first suggested as a self-healing technique for polymer matrix by White et al. [12]. White et al. suggested a healing system based on the ring-opening metathesis polymerization (ROMP) of dicyclopentadiene (DCPD) via Grubbs' catalyst. Microcapsules with urea-formaldehyde (UF) shell filled with DCPD were dispersed into epoxy matrix with Grubbs' catalyst and the addition of microcapsules and catalyst increased the toughness of the epoxy [12]. This DCPD-Grubs catalyst system and the ROMP reaction were further investigated, and the healing efficiency was shown to be related to the concentration and size of the microcapsules and the catalyst [2, 13]. The low viscosity of the DCPD and the rapid polymerization at ambient conditions are strong advantages of the system, which has been extensively studied and reported as an efficient self-healing system [2, 6, 8].

5-Ethyldiene-2-norbornene (ENB) has also been investigated as a healing agent for the capsule-based self-healing mechanism due its higher ROMP reactivity when compared with DCPD. Romero-Sabat et al. [14] used ENB/DCPD microcapsules (ENB/DCPD 80:20 wt.%) and Ru-based catalysts (Hoveyda Grubbs 2nd generation and Grubbs 3rd generation) dispersed into epoxy matrix and found healing efficiency values up by 130% at low and ultra-low temperatures [14].

Although the capsule/dispersed catalyst healing system (**Figure 3a**) was first proposed, the capsule-based mechanism has been investigated also using different encapsulated agents in a dual capsule system. In this system, the monomer and the hardener are encapsulated separately and embedded into the matrix (**Figure 3b**).

Yuan et al. [15] presented the healing efficiency of a dual capsule system with a microcapsules shell made by poly(melamine-formaldehyde) (PMF) filled with epoxy prepolymer and its hardener. Diglycidyl ether of bisphenol A(DGEBA) epoxy-loaded microcapsules and microcapsules filled with hardener (2,4,6-tris (dimethylaminomethyl)phenol) were dispersed in epoxy matrix and tapered double cantilever beam (TDCB) were used to assess the self-healing efficiency of the system. For a concentration of 20 wt.% of microcapsules (10 wt.% of epoxy-loaded and 10 wt.% of hardener-loaded microcapsules), healing efficiency higher than 90% was found under 200°C [15].

Kosarli et al. [16] proposed the use of UF microcapsules filled with epoxy resin nanomodified with Multi-Wall Carbon Nanotubes (MWCNTs) as healing agent. The conductive microcapsules aimed to recover mechanical and electrical performance of the capsule-based self-healing system. 0.1 wt% of MWCNTs was dispersed into epoxy healing agent using high shear stirrer previous to the encapsulation. UF-wall microcapsules were produced using in situ emulsification polymerization. Then, the DGEBA epoxy matrix was modified with 0.5 wt.% MWCNTs, 20 wt.% produced microcapsules, and 3 wt.% of catalyst (aluminum (III) triflate $(Al(OTf)_3)$. The self-healing efficiency was assessed through fracture toughness tests and impedance spectroscopy measurements. The capsule-based self-healing polymer presented about 22% increase on mechanical properties when compared with neat matrix. In addition, a recovery rate of more than 80% of fracture toughness and electrical properties was found [16].

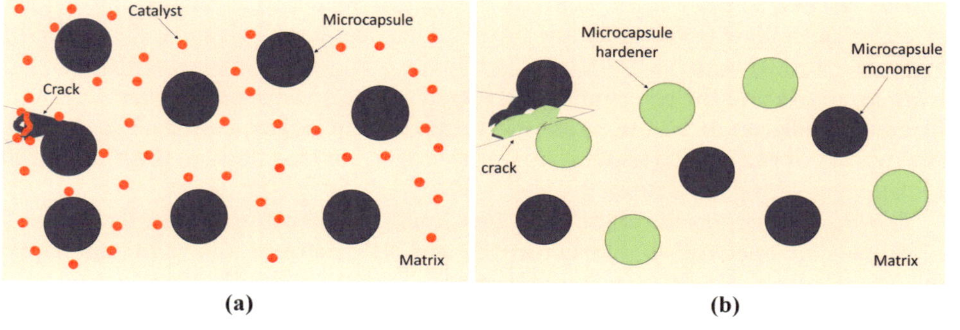

Figure 3.
Schematic of capsuled-based mechanism: a) microcapsule/disperse catalyst system and b) dual capsuled-based system.

Ebrahimnezhad-Khaljiri et al. [17] used capsuled-based extrinsic mechanism to manufacture self-healing glass fiber/epoxy composites. Microcapsules with epoxy as healing agent were first dispersed into epoxy matrix in three different concentrations: 7 wt.%, 14 wt.%, and 21 wt.% using mechanical stirrer. Then, 2 wt.% of imidazole-based catalyst ($NiCl_2(imidazole)_4$) was incorporated within the matrix. 4-ply and 12-ply glass fiber/epoxy were manufactured by hand lay-up using the modified epoxy matrix. Tensile and interlaminar shear strength (ILSS) tests were used for self-healing efficiency investigation. As a result, composites with 14 wt.% of microcapsules were presented as optimum results, considering the levels of healing efficiency for tensile and shear strength [17].

Capsule-based self-healing approaches have also been investigated for coatings and corrosion protection. For these types of coatings, despite polymerizable materials, corrosion inhibitors can be encapsulated. The release of the healing agent to the damaged areas promotes the filling of the cracks of the coating, thereby creating a protective barrier. This barrier that is formed acts as a corrosion protection, protecting the metal surface from oxidizing agents [18, 19].

For the capsule-based self-healing system, the encapsulation process is one of the key factors during the processing. Several encapsulation methods as coacervation, phase separation, in situ and interfacial polymerization, layer-by-layer assembly and meltable dispersion are used for encapsulating reactive materials. However, interfacial and in situ polymerization have been largely used as an encapsulation technique for capsule-based self-healing systems [20, 21]. The synthesis of microcapsules is a complex process and must be considered a large range of parameters. The thickness of the wall and the stiffness of the capsules must ensure that the containers remain unbroken during the processing and the rupture during cracking. The capsules must have the appropriate size and amount of healing agent without leakage. In addition, the quality of the surface is also important to guarantee good adhesion with the matrix. Synthesis parameters such as temperature, stirring rate, and pH can be adjusted to influence the size and quality of microcapsules [20, 22, 23].

Besides the encapsulation process, the incorporation of the capsules into the matrix is an important step during the process and the optimization of the dispersion of the microcapsules is a challenge. The challenges faced in the processing are considered the limitations of this self-healing mechanism. In addition, the limited amount of healing available restrains the multiple healing cycles [20]. Once the microcapsule breaks and the healing agent is polymerized, the containers cannot be refilled, and repetitive repair does not occur in the same location.

2.1.2 Vascular network self-healing

The vascular self-healing system has the healing agents encapsulated into capillaries such as hollow fibers or hollow tubes. As capsule-based systems, the capillaries can be filled with healing agents, while the catalyst is dispersed within the polymer matrix and the resin and the hardener can be encapsulated separately in different capillaries. In addition, one-part resin system is also possible (**Figure 4**). The rupture of the capillaries releases the healing agent that flows to the microcrack surfaces recovering the damaged areas.

The vascular networks can be unidimensional, bidimensional, or be formed as a 3-dimensional network-like structure. 2D and 3D structures, due to the interconnected vascular network, have the advantages of repetitive healing in the same area. In addition, in addition to working as a healing agent container, the vascular networks also act as reinforcement for the polymeric matrices [5, 24].

Williams et al. [25] used hollow glass fibers (HGF) as storage vessels for the healing agent. The HGF were infused with a two-part epoxy system and inserted

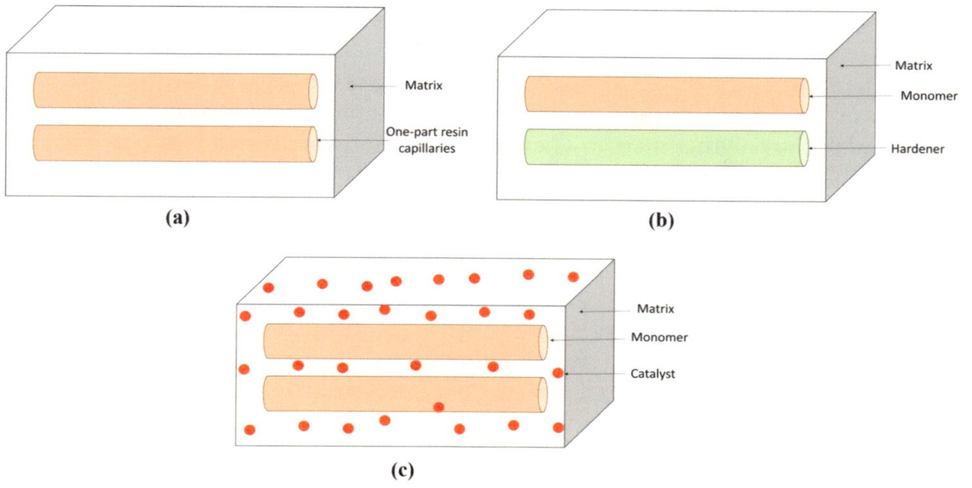

Figure 4.
a) One-part resin system, b) resin and hardener encapsulated separately and c) encapsulated resin with dispersed catalyst for vascular network self-healing system.

into carbon fiber reinforced plastic (CFRP) prior to the hand lay-up lamination. The CFRP showed the ability to restore compressive strength after low velocity impact damage [25].

Mohammadi et al. [26] investigated the self-healing efficiency of a glass fiber/epoxy composite containing microvascular channels. The one-dimensional vascular network was fabricated with the assistance of polymide wires. The polymeric wires were used as re-tractable preforms and positioned between the glass fiber during the resin impregnation via hand lay-up. The wires were covered with silicone and pulled out after the curing. The hollow channels were then filled with the healing agent. A three-part healing agent consisting of epoxy resin, maleic anhydride hardener, and CuBr2 (2-Methylimidazole) accelerator was used to fill the microvascular channels. After damage, the healing was activated by heating at 130°C for 30 min. Healing efficiencies above 50% were reported under tensile and flexural loadings [26].

The DCPD/Grubbs' catalysts system is also largely used as a healing agent for hollow fiber/capillaries vascular networks. Radovic et al. [27] investigated the influence of the solvent on the stability of the Grubbs' catalyst and the healing efficiency of the glass fiber/epoxy composites with embedded filled glass capillaries. Dichloromethane (DCM) and toluene (TO) were used for the dissolution of the Grubbs' catalyst. A two-parts glass capillaries system was used: monomer and catalyst encapsulated in different capillaries. The monomer was a solution of 10 wt.% DCPD in dimethylformamamide 99.8% (DMF) and the catalyst was a solution of 1 wt.% of Grubbs catalyst dissolved in two different solvents (DCM e TO) for comparison. Controlled energy impact tests showed that samples with DCM solvent presented 60% of recovery of impact strength after healing, while samples with TO solvent exhibited 51% of recovery. In addition, results showed that the filled capillaries for healing purpose also are suitable as reinforcement for the composite during loading [27].

Extrinsic self-healing mechanisms have attracted attention and been extensively studied through the last decades. The main advantage of this type of mechanism is that it does not require external stimulus, it being a totally autonomic self-healing mechanism. However, repeatable healing is limited due to the limited amount of healing agent available.

2.2 Intrinsic self-healing

The concept of intrinsic self-healing mechanism is considered more competitive in terms of manufacturing and applications because the intrinsic healing ability is conferred to the polymer matrix. Thermoset polymer matrices with intrinsic healing capability have the advantage of multiple healing cycles as long as damage to the reinforcement fibers is not excessive or highly localized. Different mechanisms have been proposed in the last decade for the production of self-healing polymer matrices. The healing capability is achieved through the modification of the polymeric matrix to contain intrinsic functionality or healing agent [28, 29].

Intrinsic self-healing systems are obtained through the modification of the structure of polymeric matrix (crosslinking density, aromatic chains, crystallinity, for instance) or by the modification of the polymeric matrix through the addition of a thermoplastic modifier, obtained by a miscible or immiscible thermoplastic/ thermoset blend [28].

2.2.1 Reversible polymer networks self-healing system

For this intrinsic mechanism, the polymeric matrix is chemically modified with functional groups that experience reversible reactions in the presence of external stimulus. Then, the healing is achieved due to the resulting conformation changes of the dynamic dissociation and re-association of stress-bearing bonds. Different types of dynamic covalent bonds and non-covalent bonds (referred to a supramolecular interactions) have been investigated for self-healing purposes such as Diels-Alder, disulfide exchange reactions, hydrogen bonds, and poly(dimethylsiloxane) reactions. The healing is usually initiated by temperature, although studies showing pH changes, chemical additives and irradiation as stimulus for the healing activation can be found [29, 30].

Diels-Alder and hydrogen binding active groups are extensively investigated to produce polymer networks with self-healing ability due to thermally reversible properties. In these networks, the bonds reconnection that allow the repair, can occur under different temperatures and conditions. The introduced functional groups are decisive for the healing event and the ratio between these groups can control the conditions necessary for the healing activation; that is, the levels of energy necessary to trigger the reversible interactions [31]. In addition, the introduced functional groups are decisive for the healing efficiency, as characteristics such as association and flexibility and architecture of the molecule have to be considered and they can influence the success of healing [29].

Ehrhardt et al. synthesized a UV-cured reversible poly(methacrylate) network capable of healing damage at ambient conditions. Reversibility is introduced by Diels-Alder bonds formed through the reaction between maleimide and furan. UV-cured polymer films were ground into a powder, which was further compressed in a mold in order to bring off particles into close contact. Compressed samples were healed for 7 days at 1 bar and 20°C. As result, the network was healed in the fully vitrified state [32].

Grande et al. [33] investigated the healing behavior of aromatic disulphide-based poly(urea-urethane) (PUU). Different combinations of disulphide and polyurethane pre-polymers were used to obtain different crosslink densities with the same number of reversible covalent bonds. Then, the effect of the polymer structure was investigated. Dynamical mechanical thermal analysis (DMTA), tensile tests and single edge notch tensile (SENT) fracture tests were carried out to assess the self-healing efficiency at room temperature. Results suggested that the structure of the network interferes with the levels of recovery after damage. The addition of high contents of difunctional units resulted in polymers with low tensile strength but high viscoelastic properties and healing efficiency [33].

Supramolecular interactions easily dissociate under mechanical loading due the weak nature of the interactions. Nevertheless, these bonds can re-associate with minimal external trigger. Intrinsic self-healing systems based on supramolecular interactions present as major benefit the ability to self-heal at low temperatures, near to room temperature [34]. However, the dynamic nature of the reversible bonds implies in temperature-dependent behavior, and the long-term mechanical stability is an issue. Polymers network with covalent reversible bonds, on the other hand, potentially are able to operate in a higher temperature range, presenting more mechanical properties stability, but require higher temperatures for healing activation. In addition, the design of reversible polymer networks can involve long, expensive, and complex synthesis steps [33, 34].

2.2.2 Thermoplastic/thermoset blends self-healing system

The addition of thermoplastic modifiers is considered promising as an intrinsic healing mechanism proposed for thermoset resin. This mechanism is based on the incorporation of a solid-state thermoplastic healing agent in the thermoset matrix. During heating, the movement of these molecules promotes the flow of the healing agent into the damaged regions, filling the crack and providing the healing of the thermoset matrix [8, 35].

Three properties of thermoplastics are considered critical for the use of these polymers as the healing agent: fluidity, reactivity of functional groups, and adhesive properties. The thermoplastic must have a melting point and viscosity that provide flow inside the cracks besides being chemically reactive functional groups with the groups of the thermoset matrix. The thermoplastic must also have good adhesive properties, providing strong bonding with the matrix during crack repair [8].

In addition, as for this mechanism, the external activation to promote the healing is necessary, selecting the optimal parameters to promote the flow of the thermoplastic healing agent as an important step. Temperature and pressure are usually applied to enable the healing. In terms of temperature, the activation temperature has to be enough to provide the flow of the thermoplastic, the temperature being above its melting point. In practice, is also suggested that the activation of the healing occurs at temperatures above the glass transition of the matrix. The mobility of the crosslink matrix can increase the diffusion between thermoplastic and matrix and facilitate the adhesion of the healing agent to the matrix networks [35, 36].

Considering these requirements, some thermoplastic polymers have been suggested as possible healing agents for epoxy resin. Poly bisphenol-A-co-epichlorohydrin, polycaprolactone (PCL), poly (methyl methacrylate) (PMMA), poly (ethylene-co-methyl acrylate) (EMA), poly (ethylene-co-methacrylic acid) (EMAA), and poly (ethylene glycidyl methacrylate) (EGMA) are examples of thermoplastic polymers studied as healing agents for epoxy resin [6, 8].

Varley et al. [35] investigated the healing efficiency of six different thermoplastic modifiers dispersed into epoxy amine network. Ethylene vinyl acetate (EVA), poly(vinyl-butyral) (PVB), styrene-ethylene-butadiene copolymer (SEBS), poly(ethylene-co-glycidyl)-methacrylate (PEGMA), polyethylene-co-methacrylic acid (EMAA), and acrylonitrile-butadiene-styrene (ABS) were used. The healing agents were cryogenically ground and the particles (with size distribution between 149 and 295 μm) were blended within epoxy resin. The healing efficiency was assessed through single end notched beam (SENB) tests and four healing cycles were performed in order to investigate the ability for repetitive healing. All healing agents presented load recovery after the first healing cycle. However, ABS, SEBS, and PVB showed lower repeatability of healing. EMAA, PEGMA, and EVA presented high levels of mendable behavior, especially after further healing cycles [35].

The poly (ethylene-co-methacrylic acid) (EMAA) has been extensively studied as a healing agent for epoxy resin and FRP, presenting high healing efficiency. In the case of EMAA, the thermoplastic is incorporated to the thermoset matrix and remains insoluble. When used as a healing agent in high concentrations (concentrations above 5 wt.% are usually reported), EMAA is considered effective in healing the cracks in both epoxy resin and carbon fiber epoxy composites [37]. EMAA was first suggested as a healing agent for epoxy networks by Meure et al. [38]. EMAA particles with concentration of 15 wt.% were incorporated into epoxy resin and the healing efficiency was assessed by TDCB and SENB tests. As a result, a strong recovery due to the thermal expansion of the EMAA and the formation of an adhesive layers between adjacent damaged surfaces was found. Further studies [39, 40] showed that strong chemical interactions between EMAA and epoxy networks are responsible for the healing mechanism [38–40].

EMAA was also reported as an efficient self-healing agent for CFRC, exhibiting more than 100% recovery of properties [41, 42]. For FRC, the self-healing efficiency of the system is a result of the plastic deformation of the EMAA during the heating, which gets transformed into thin ligaments, which bridge the crack. These thermoplastic ligaments are able to transfer the applied tension through the crack, increasing the interlaminar resistance of the system [43].

Even though temperature is highly used to activate EMAA as a healing agent, Hargou et al. [44] showed that ultrasonic welding can be used to activate the healing mechanism for carbon fiber/epoxy systems. The ultrasonic vibration generated the temperature to activate the healing [44].

Miscible blends of epoxy cured with 4,4′-diaminodiphenylsulfone(DDS) and PCL as thermoplastic healing agent have also been investigated. Cohades et al. [45] used epoxy-PCL blends with 25 vol.% of PCL to manufacture glass fiber reinforced composites and the self-healing ability was assessed through compression after impact (CAI) tests. Results showed that even though epoxy-PCL composites presented a lower resistance to crack propagation when compared with pure epoxy composites, the PCL modified composites presented a recovery rate of 22 to 96% of compressive residual strength after low-velocity impact damage. Low levels of recovery were ascribed to fiber damage, which cannot be healed by the self-healing mechanism focused on the healing of matrix microcracking [45].

Kong et al. [46] incorporated fusible thermoplastic polystyrene (PS) as a healing agent into a shape of memory polyimide matrix. Shapes of memory polymers (SMPs) have the ability to return to the original shapes under external stimuli such as heat and light. Then, the combination of self-healing performance with the shape of memory effect through self-healing SMPs have been studied. In their study, Kong et al. incorporated 8% of PS into the shape of memory polyimide matrix and showed that the combined effect can heal damages such as cracks, pierced holes, and cuts at 243°C [46].

Intrinsic self-healing mechanisms have as its main advantage the indefinite repeatability, since repair is inherent to the chemical structure of the polymer. However, the need for some external activation can be considered a limitation of this repair mechanism and the design of the requirements for healing activation is specific for each system [47]. In addition, due to the need of modification of the matrix network, synthesizing intrinsic self-healing polymers to obtain high healing efficiency without mechanical strength loss is a challenge.

3. Conclusion

Self-healing materials are able to heal themselves when damaged, restoring initial properties fully or partially. Inspired by natural mechanisms, self-healing mechanisms have been explored for all types of materials, including metals, polymers, ceramics, and composites.

In the last decades, great progress has been made in the design and development of self-healing polymeric materials and the existing self-healing polymers are considered smart materials able to recover their properties autonomically or non-autonomically. These materials have been suggested for a series of applications such as water membranes, lithium batteries, energy transducers, biomedical devices, flexible electronics, and soft robotics [48].

Ideal self-healing polymers cannot significantly decrease the initial properties of the virgin material. In addition, the methods to assess the healing agent efficiency still need to be better standardized.

Self-healing materials are a new class of smart materials and have attracted a significant interest for research. Several self-healing techniques have been developed and investigated in the last decades. All techniques developed are reported as promising but they also presented with limitations and downsides. Adjustments are still necessary in order to apply the techniques to industrial scale production. Thus, further investigations and research are still necessary.

Conflict of interest

The authors declare no conflict of interest.

Author details

Allana Azevedo do Nascimento
Department of Materials Science and Engineering, Federal University of Rio Grande do Norte, Natal, Brazil

*Address all correspondence to: allanaazevedon@hotmail.com

References

[1] Wu DY, Meure S, Solomon D. Self-healing polymeric materials: A review of recent developments. Progress in Polymer Science 2008;**33**(5):479-522.

[2] Kessler MR. Self-healing: A new paradigm in materials design. Proceedings of the Institution of Mechanical Engineers - Part G: Journal of Aerospace Engineering 2007;**221**(4): 479-495.

[3] Tan PS, Somashekar AA, Casari P, Bhattacharyya D. Healing efficiency characterization of self-repairing polymer composites based on damage continuum mechanics. Composite Structures. 2019;**208**(September 2018):367-376. Available from: https://doi.org/10.1016/j.compstruct.2018.09.091

[4] Scheiner M, Dickens TJ, Okoli O. Progress towards self-healing polymers for composite structural applications. Polymer (Guildf). 2016;**83**:260-282. Available from: http://dx.doi.org/10.1016/j.polymer.2015.11.008

[5] JE PC, Sultan MT, Selvan CP, Irulappasamy S, Mustapha F, Basri AA, et al. Manufacturing challenges in self-healing technology for polymer composites — A review. Journal of Materials Research and Technology. 2020;**9**(4):7370-7379. Available from: https://doi.org/10.1016/j.jmrt.2020.04.082

[6] Blaiszik BJ, Kramer SLB, Olugebefola SC, Moore JS, Sottos NR, White SR. Self-healing polymers and composites. Annual Review of Materials Research 2010;**40**:179-211.

[7] Reddy KR, El-Zein A, Airey DW, Alonso-Marroquin F, Schubel P, Manalo A. Self-healing polymers: Synthesis methods and applications. Nano-Structures & Nano-Objects. 2020;**23**:100500. Available from: https://doi.org/10.1016/j.nanoso.2020.100500

[8] Hia IL, Vahedi V, Pasbakhsh P. Self-healing polymer composites: Prospects, challenges, and applications. Polymer Reviews. 2016;**56**(2):225-261. Available from: http://dx.doi.org/10.1080/15583724.2015.1106555

[9] Bekas DG, Tsirka K, Baltzis D, Paipetis AS. Self-healing materials: A review of advances in materials, evaluation, characterization and monitoring techniques. Composites Part B: Engineering 2016;**87**:92-119. Available from: http://dx.doi.org/10.1016/j.compositesb.2015.09.057

[10] Peñas-Caballero M, Hernández Santana M, Verdejo R, Lopez-Manchado MA. Measuring self-healing in epoxy matrices: The need for standard conditions. Reactive and Functional Polymers 2021;**161** (February):104847

[11] Paolillo S, Bose RK, Santana MH, Grande AM. Intrinsic self-healing epoxies in polymer matrix composites (Pmcs) for aerospace applications. Polymers (Basel). 2021;**13**(2):1-32.

[12] White SR, Sottos NR, Geubelle PH, Moore JS, Kessler MR, Sriram SR, et al. Autonomic healing of polymer composites. Nature. 2001;**409**:794-817.

[13] Brown EN, Sottos NR, White SR. Fracture testing of a self-healing polymer composite. Experimental Mechanics 2002;**42**(4):372-379.

[14] RRomero-Sabat G, Gago-Benedí E, Roa Rovira JJ, González-Gálvez D, Mateo A, Medel S, et al. Development of a highly efficient extrinsic and autonomous self-healing polymeric system at low and ultra-low temperatures for high-performance

applications. Composites Part A: Applied Science and Manufacturing 2021;**145**:106335

[15] Yuan YC, Ye XJ, Rong MZ, Zhang MQ, Yang GC, Zhao JQ. Self-healing epoxy composite with heat-resistant healant. ACS Applied Materials & Interfaces 2011;**3**(11):4487-4495.

[16] Kosarli M, Foteinidis G, Tsirka K, Bekas DG, Paipetis AS. Concurrent recovery of mechanical and electrical properties in nanomodified capsule-based self-healing epoxies. Polymer (Guildf). 2021;**227**:123843. Available from: https://doi.org/10.1016/j.polymer.2021.123843

[17] Ebrahimnezhad-Khaljiri H, Eslami-Farsani R. The tensile properties and interlaminar shear strength of microcapsules-glass fibers/epoxy self-healable composites. Engineering Fracture Mechanics. 2020;**230**:106937. Available from: https://doi.org/10.1016/j.engfracmech.2020.106937

[18] Liu T, Ma L, Wang X, Wang J, Qian H, Zhang D, et al. Self-healing corrosion protective coatings based on micro/nanocarriers: A review. Corrosion Communications. 2021;**1**:18-25. Available from: https://doi.org/10.1016/j.corcom.2021.05.004

[19] An S, Lee MW, Yarin AL, Yoon SS. A review on corrosion-protective extrinsic self-healing: Comparison of microcapsule-based systems and those based on core-shell vascular networks. Chemical Engineering Journal. 2018;**344**:206-220. Available from: https://doi.org/10.1016/j.cej.2018.03.040

[20] Islam S, Bhat G. Progress and challenges in self-healing composite materials. Mater Adv. 2021;**2**(6):1896-1926.

[21] MMarinho TMA, Pontes VS, Melo JDD, Costa MCB. Evaluation of the synthesis variables' effect on the properties of PUF/ENB microcapsules using experimental design. Journal of Polymer Research. 2021;**8**:136. Available from: https://doi.org/10.1007/s10965-021-02494-1

[22] de Carvalho ACM, Ferreira EP da C, Bomio M, Melo JDD, Cysne Barbosa AP, Costa MCB. Influence of synthesis parameters on properties and characteristics of poly (urea-formaldehyde) microcapsules for self-healing applications. Journal of Microencapsulation. 2019;**36**(4):410-419. Available from: https://doi.org/10.1080/02652048.2019.1638462

[23] Kosarli M, Bekas D, Tsirka K, Paipetis AS. Capsuled-based self-healing polymers and composites. Thomas S, Surend A, editors. Self-Healing Polymer-Based Systems. Elsevier; 2020. 259-278. http://dx.doi.org/10.1016/B978-0-12-818450-9.00010-6

[24] Lee MW, An S, Yoon SS, Yarin AL. Advances in self-healing materials based on vascular networks with mechanical self-repair characteristics. Advances in Colloid and Interface Science. 2018;**252**:21-37. Available from: http://dx.doi.org/10.1016/j.cis.2017.12.010

[25] Williams GJ, Bond IP, Trask RS. Compression after impact assessment of self-healing CFRP. Composites Part A: Applied Science and Manufacturing. 2009;**40**:1399-1406. Available from: http://dx.doi.org/10.1016/j.compositesa.2008.05.021

[26] Mohammadi MA, Eslami-Farsani R, Ebrahimnezhad-Khaljiri H. Experimental investigation of the healing properties of the microvascular channels-based self-healing glass fibers/epoxy composites containing the three-part healant. Polymer Testing. 2020;**91**:106862. Available from: https://doi.org/10.1016/j.polymertesting.2020.106862

[27] Radovic I, Stajcic A, Radisavljevic A, Veljkovic F, Cebela M, Mitic V V., et al. Solvent effects on structural changes in self-healing epoxy composites. Materials Chemistry and Physics. 2020;**256**: 123761. Available from: https://doi.org/10.1016/j.matchemphys.2020.123761

[28] Cohades A, Manfredi E, Plummer CJG, Michaud V. Thermal mending in immiscible poly (ϵ-caprolactone)/epoxy blends. European Polymer Journal 2016;**81**: 114-128.

[29] Garcia SJ. Effect of polymer architecture on the intrinsic self-healing character of polymers. European Polymer Journal 2014;**53**:118-125. Available from: http://dx.doi.org/10.1016/j.eurpolymj.2014.01.026

[30] Hornat CC, Urban MW. Shape memory effects in self-healing polymers. Progress in Polymer Science. 2020;**102**:101208. Available from: https://doi.org/10.1016/j.progpolymsci.2020.101208

[31] Araya-Hermosilla R, Lima GMR, Raffa P, Fortunato G, Pucci A, Flores ME, et al. Intrinsic self-healing thermoset through covalent and hydrogen bonding interactions. European Polymer Journal 2016;**81**:186-197. Available from: http://dx.doi.org/10.1016/j.eurpolymj.2016.06.004

[32] Ehrhardt D, Van Durme K, Jansen JFGA, Van Mele B, Van den Brande N. Self-healing UV-curable polymer network with reversible Diels-Alder bonds for applications in ambient conditions. Polymer (Guildf). 2020;**203**:122762. Available from: https://doi.org/10.1016/j.polymer.2020.122762

[33] Grande AM, Martin R, Odriozola I, van der Zwaag S, Garcia SJ. Effect of the polymer structure on the viscoelastic and interfacial healing behaviour of poly(urea-urethane) networks containing aromatic disulphides. European Polymer Journal. 2017;**97**:120-128. Available from: https://doi.org/10.1016/j.eurpolymj.2017.10.007

[34] Gadwal I. A brief overview on preparation of self-healing polymers and coatings via hydrogen bonding interactions. Macromol. 2020;**1**:18-36.

[35] Varley RJ, Craze DA, Mouritz AP, Wang CH. Thermoplastic healing in epoxy networks: Exploring performance and mechanism of alternative healing agents. Macromolecular Materials and Engineering 2013;**298**(11): 1232-1242.

[36] Pingkarawat K, Dell'Olio C, Varley RJ, Mouritz AP. Poly(ethylene-co-methacrylic acid) (EMAA) as an efficient healing agent for high performance epoxy networks using diglycidyl ether of bisphenol A (DGEBA). Polymer (Guildf). 2016;**92**:153-163. Available from: http://dx.doi.org/10.1016/j.polymer.2016.03.054

[37] Pingkarawat K, Wang CH, Varley RJ, Mouritz AP. Mechanical properties of mendable composites containing self-healing thermoplastic agents. Composites Part A: Applied Science and Manufacturing 2014;**65**:10-18. Available from: http://dx.doi.org/10.1016/j.compositesa.2014.05.015

[38] Meure S, Wu DY, Furman S. Polyethylene-co-methacrylic acid healing agents for mendable epoxy resins. Acta Mater 2009;**57**(14):4312-4320. Available from: http://dx.doi.org/10.1016/j.actamat.2009.05.032

[39] Meure S, Varley RJ, Wu DY, Mayo S, Nairn K, Furman S. Confirmation of the healing mechanism in a mendable EMAA-epoxy resin. European Polymer Journal 2012;**48**(3):524-531. Available from: http://dx.doi.org/10.1016/j.eurpolymj.2011.11.021

[40] Meure S, Wu DY, Furman SA. FTIR study of bonding between a thermoplastic healing agent and a mendable epoxy resin. Vibrational Spectroscopy 2010;**52**(1):10-15.

[41] do Nascimento AA, Fernandez F, da Silva FS, Ferreira EP.C., José JD, Cysne Barbosa AP. Addition of poly (ethylene-co-methacrylic acid) (EMAA) as self-healing agent to carbon-epoxy composites. Composites Part A: Applied Science and Manufacturing. 2020;**137**(February):106016. Available from: https://doi.org/10.1016/j.compositesa.2020.106016

[42] Meure S, Furman S, Khor S. Poly[ethylene-co-(methacrylic acid)] healing agents for mendable carbon fiber laminates. Macromolecular Materials and Engineering 2010;**295**(5):420-424.

[43] Pingkarawat K, Bhat T, Craze DA, Wang CH, Varley RJ, Mouritz AP. Healing of carbon fibre-epoxy composites using thermoplastic additives. Polymer Chemistry 2013;**4**(18):5007-5015.

[44] Hargou K, Pingkarawat K, Mouritz AP, Wang CH. Ultrasonic activation of mendable polymer for self-healing carbon-epoxy laminates. Composites Part B: Engineering 2013;**45**(1):1031-1039. Available from: http://dx.doi.org/10.1016/j.compositesb.2012.07.016

[45] Cohades A, Michaud V. Damage recovery after impact in E-glass reinforced poly(ε-caprolactone)/epoxy blends. Composite Structures 2017;**180**:439-447.

[46] Kong D, Li J, Guo A, Zhang X, Xiao X. Self-healing high temperature shape memory polymer. European Polymer Journal. 2019;**120**:109279. Available from: https://doi.org/10.1016/j.eurpolymj.2019.109279

[47] Varley RJ, Charve F. EMAA as a healing agent for mendable high temperature epoxy amine thermosets. Composites Part A: Applied Science and Manufacturing 2012;**43**(7):1073-1080. Available from: http://dx.doi.org/10.1016/j.compositesa.2012.01.018

[48] Almutairi MD, Aria AI, Thakur VK, Khan MA. Self-healing mechanisms for 3D-printed polymeric structures: From lab to reality. Polymers (Basel). 2020;**12**(7):1-27.

Chapter 2

Design, Simulation, and Analysis of the Extrusion Process of a PVC Thermoplastic Profile to Optimize the Design of the Die and the Machine Parameters

Carlos José Salvador Tomassini

Abstract

The objective of this work is to verify the design of an existing die for the manufacture of an extruded profile using the simulation of the flow in the head using a simulation software that uses computational fluid dynamics and also the experimental design and construction of a calibrator by means of the extrusion the geometry and desired dimensions of the profile. The rheological behavior of rigid PVC in the extruded molten state was investigated, which in itself is a difficult target due to the intrinsic weakness of this polymer that degrades when heated above 140°C. By means of a special capillary rheometer, rheological data, k and n of the power law, were obtained to introduce them, together with the process input parameters and the flow channel geometry in the simulation software. The flow channel was drawn with the head and calibrator using CAD-3D software. The different parts of the calibrator were manufactured and assembled into the equipment. The extrusion was performed with the process parameters: screw speed and material temperature used in the simulation software. The results obtained by the extrusion, geometry and final dimensions of the profile, mass flow, pressure, and temperature in the head were compared with those delivered by the software, being the same satisfactory.

Keywords: simulation, extrusion, PVC (polyvinyl chloride), process, design, rheology

1. Introduction

Extrusion molding simulation allows you to avoid or reduce the high costs of the manufacture of the die or head and the probability of avoiding problems during production. On the other hand, costly modifications are saved, eliminating the current trial-and-error process, which added to the loss of time and lost profits, making simulation a tool of great importance for the national industry, especially for small- and medium-sized companies, since computer tools are not used in the design of heads and in the extrusion process of thermoplastic profiles in our country. Computer simulation, using specific software, allows to accelerate these processes, making possible all the necessary considerations in the design before building the head, especially for processes and design of profiles with complicated geometries and even with the use of different types of materials to manufacture the same. In addition, it is also important to know how to design a caliper in order to obtain an extruded profile with the desired final geometry and dimensions [1].

As antecedents of the international research works that were carried out on the subject, some of them can be cited, although there are differences with respect to the present work, and in the materials, simulation software and extruder equipment, including the ones used in the final process of calibration and cooling of the profile, are the ones that most resemble it.

Therefore, we can cite the one carried out by Srinivasa et al. [2], in which the simulation is performed on an existing head with Polyflow software that uses computational fluid dynamics, with a double screw extruder and using polystyrene as a material, which is simpler to process than the composite of rigid PVC that uses the reverse extrusion simulation process with the objective that the results of the same are used to improve the design of a new head. In the conclusion, it is commented that discrepancies were found between the results of the computational simulation and the extrusion in the dimensions of the profile between 5 and 10%.

On the other hand, the work of Gupta [2] can be cited, who uses the finite element method for a three-dimensional simulation applied to the flow of the molten polymer and analyzes the shear and elongation viscosity in the head in a final rectangular way, which is represented by the truncated power law model. In this work, he obtains a recirculation of the molten material at the vertices and a loss of pressure, which causes an abrupt contraction that increases rapidly with the exponent n of the power law. Therefore, it manages to avoid the said recirculation and pressure loss, so that the contraction in the final profile is less pronounced.

Finally, Bogale [3] in his thesis work on design and simulation of a head for a rectangular profile, through modeling and simulation using the COMSOL Multiphysic software and the CA-RREAU model [4], also uses the low-density polyethylene as a study material and with a temperature of 220°C in the head. With the characteristics of the extruder screw, the flow rate is calculated, with which the pressure in the extruder could be obtained, since when it is minimum the output flow is maximum and vice versa. Although the design of the head was optimized in order to obtain a homogeneous distribution of the shear stress in the cross section of the same, it was not possible to simulate the exact dimensions of the head for a rectangular profile with the Comsol software.

In addition, this research aims to evaluate the predictive capacity of the Polyflow simulation software that uses computational fluid dynamics. Zhao [5] applied to the case of a rigid PVC profile, which has certain particular characteristics such as the aforementioned thermal sensitivity.

Thus, from the simulation, the contour values of the profile at the head outlet, the distribution of speeds that influence the contractions and stresses in the final profile and also the deformations and shear stresses that determine the physical and mechanical properties of the final profile, are obtained. In addition, the extrusion of the profile is carried out, at a certain RPM of the screw, obtaining the values of the same from the pressure and temperature sensors in the head, and on the other hand, the mass flow at the outlet of the head is measured. Subsequently, the results of both processes are compared and the results that are described in detail below in the corresponding item are obtained.

2. Materials and methods

2.1 Material

2.1.1 Characteristics of material:

The material under investigation is a rigid PVC compound from the company ALFAVINIL RE97/1-773, which is currently used widely in the manufacture of all types of profiles for construction, in the refrigeration industry, among other applications, due to its excellent properties of resistance to corrosive media, good thermal, electrical and acoustic insulation, low weight, and mechanical properties suitable for each use. This is possible due to the possibility of designing its formulation with different types of materials (resin, plasticizer, additives, mineral fillers, pigments, thermal stabilizers, and others). It is also an ecological material, since it has the lowest carbon footprint and can be recycled.

On the other hand, due to its chemical composition, it is a thermally sensitive material, degrading from 140°C; therefore, for its processing, in addition to containing a thermal stabilizer in its composition, the use of special extruders is required, with screws with a low compression ratio (2:1), which perform the plasticization of the molten material with a low shear deformation or also known as shear rate, so that the heat provided for this process comes mainly from the heating bands of the cylinder and that the movement of the screw is mainly to improve the mixing and homogenization of the different components of the compound without generating heat by friction between the material, screw, and cylinder. Due mentoned reason, twin screw extruders, which have the best of these mentioned characteristics are recommended for use for this material, especially for medium and high productions.

2.2 Rheological characterization of the material

Most of the molten polymers behave as non-Newtonian fluids; that is to say, their shear viscosity (η), at a constant temperature, depends on the shear deformation or shear rate (γ), being able to define it (Equation 1) is the quotient between shear stress (τ) and shear rate (γ) Vlachopoulos [6].

$$\eta = \tau/\gamma \tag{1}$$

On the other hand, thermoplastics, especially PVC, is classified as pseudoplastics; that is, the viscosity decreases with shear deformation (e.g., that is exerted by the extruder screw). To carry out computer simulation, precise information on the behavior of the material under the processing conditions must be possessed: temperature, flow, density, and the rheological characteristics of the material, especially these are very important because it is a composite. Regarding the rheological characteristics, it is necessary to know the K and n values of the model to be used and that in the case of this work, the value of the shear deformations that occur in the last zone of the extruder and inside the head would apply the Power or Ostwald's Law (Eq. (2)). To obtain the values of the mentioned variables, a special capillary rheometer connected to an extruder equipment, Limper and Fattmann [7], is used, and from this test, the curve deformations of cuts versus viscosities for six conditions are obtained, as shown in **Table 1** and it is carried out by the company ALFAVINIL S.A., manufacturers of PVC composites, with a RheoDrive 7 capillary rheometer, model: Rheomex 19/25 OS, in line with an

Material	Percent (%)	Details
Resin content	66	Supply: Solvay Argentina
Plasticizer content	2	Vegetal origin, co-stabilizer
Stabilizer content	2	Stabilizer pack—lead tribasic sulfate-based lubricant
Contents of other elements	24	Micronized-coated calcium carbonate, average particle size 1.6 μm
Titanium dioxide	6	High opacity

Table 1.
Rheological values of the rigid PVC compound of ALFAVINIL code: RE97/1-773.

extruder, whose screw has a compression ratio of 2:1, L/D: 25 and a slit capillary matrix H: 2.0 mm W: 20 mm and with a material temperature profile of 150, 170, and 190°C in the extruder and 200°C in the head.

$$\eta = K.(\gamma n - 1) \qquad (2)$$

A practical way to identify a PVC compound is through its *K* value, which is the relative viscosity index [4] and which in our case has a value of 65; it is determined by testing a solution with a concentration of 5 g/L. PVC-U in cyclohexanone measures the time of passage through a capillary containing the Ostwald viscometer. This test was also carried out by the company ALFAVINIL S.A. Most of the molten polymers behave as non-Newtonian fluids, that is, their shear viscosity (η), at a constant temperature, depending on the shear deformation (γ), being able to define it (Equation 1) is the quotient between shear stress (τ) and shear rate (γ). Thermoplastics, especially PVC, is classified as pseudoplastics; that is, their shear viscosity decreases with increasing shear deformation (γ). To carry out a simulation, you must have precise information on the behavior of the material under the following processing conditions: temperature, flow, density, and rheological characteristics. Due to the fact that it is desired to model the extrusion process in the final part of the screw and in the head, the flow of the molten material has low speeds and shear deformations (γ), the latter are of the order of 50–100 1/s, For this reason, the Ostwald model or Power Law (Eq. (2)) is applied for which the k and n values need to be known, which are obtained from the rheological test. $\eta = k.\gamma\ (n - 1)$ (2). The value n is the exponent of the power law and is defined as the relationship between the stress and the rate of deformation or shear strain (γ), and the value k is the viscosity at shear rate or speed of deformation = 0 [3]. The rheological test is carried out with a special capillary rheometer connected to an extruder [8] and the values of the shear deformations vs. viscosities are obtained for six points or measurement conditions (**Table 2**).

2.3 Methods

2.3.1 Drawing of the flow channel in the head

The existing head flow channel is drawn using 3D solid Computer Aided Design software (**Figure 1**).

2.3.2 Simulation

The flow channel geometry drawing is imported into the head using the ANSYS Polyflow simulation software.

Measurement items	Shear rate (ɣ) 1/s	Shear tensile (T) bar	Viscosity (η) Pa	Mass flow (Qm) g/h	Volumetric flow (Qv) cm^3/s
1	22.7	0.7	2933.7	1730.4	0.302
2	27.5	0.8	2725.4	2101.2	0.367
3	38.6	1.0	2511.2	2944.4	0.514
4	51.6	1.2	2329.8	3940.8	0.688
5	69.8	1.4	2069.3	5324.8	0.930
6	90.4	1.7	1881.5	6899.2	1.205

Table 2.
Rheological values obtained from the rigid PVC test—ALFAVINIL code: RE97/1-773.

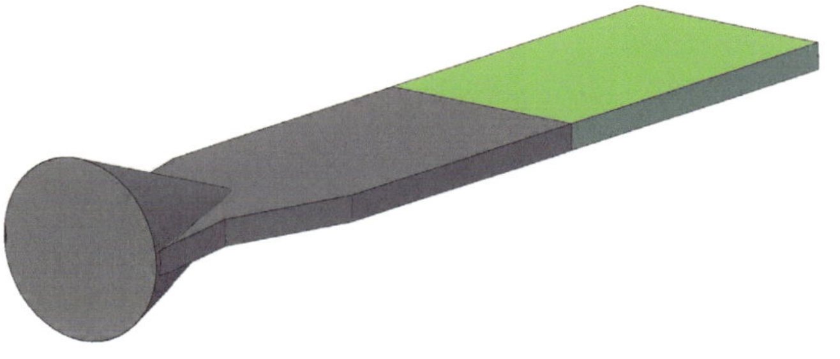

Figure 1.
Geometry flow interior head and outlet distance 60 mm (green).

Geometry meshing (**Figure 2**) is performed using the Polyflow default system, which performs the automatic subdivision into anisotropic tetrahedral elements for three-dimensional flows that have free surfaces such as the one presented in our case (flow of molten material at the head outlet).

In the same module, the names of each subdomain are incorporated, which are the areas of the melt inside the head: SD1 (**Figure 3**) and outside of it: SD2 (**Figure 4**) so that the simulation software recognizes which is each zone for the later calculations.

The names of the contours (surfaces) of each subdomain SD1_BS2 (**Figure 5**) and SD2_BS2 (**Figure 6**) are defined; the flow input surfaces to the head: SD1_BS1 (**Figure 7**) and the flow output surfaces of the same: SD2_BS1 (**Figure 8**) are also defined.

Subsequently, the setup module is entered and the input variables are entered using the Polydata program: flow (8.5 kg/h), material density (1.42 g/cm^3), and the rheological values k = 7645 and n = 0.69 of the Power Law, also defining the boundary conditions, that is, the values of speed, voltages, input, and output flows among other parameters of each subdomain and contours (surfaces). In addition, in this step, the zone in which the meshing must be performed again is defined in the programming, which is the exit zone of the molten flow from the head (see green region in **Figure 1**), so that the geometry of the profile in the molten state with the deformations that are the product of the swelling effect of the melt [3]. This effect occurs when a non-Newtonian fluid (polymer in viscoelastic molten state) passes through a restricted area, such as the internal geometry of the flow channel in the head, tensions are generated in its interior and when leaving said matrix, the material relaxes, decreasing its length and increasing its cross section. To determine the magnitude of this geometric variation of the flow at the outlet of the head, ANSYS Polyflow uses a technique called overlock optimization, which uses a

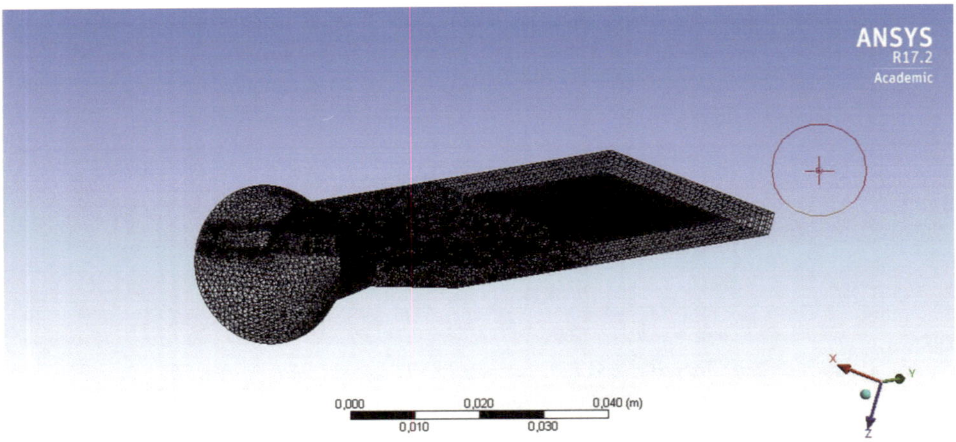

Figure 2.
Mesh of the flow channel inside the head.

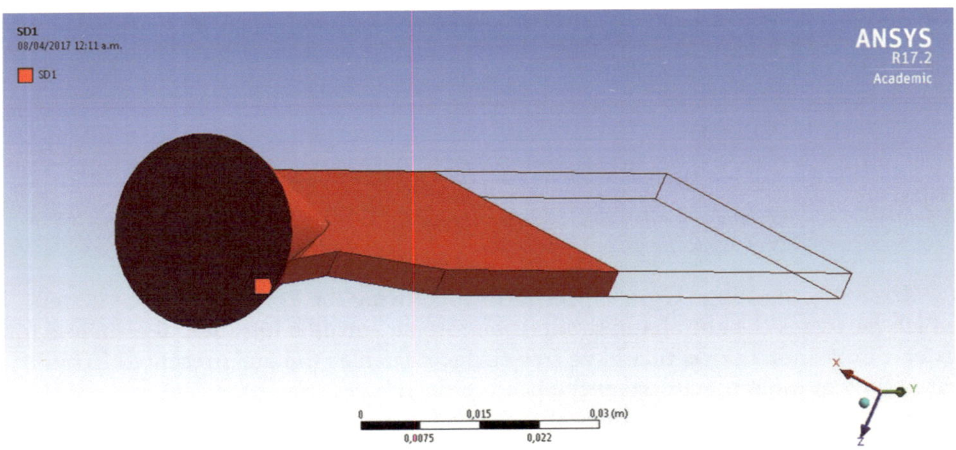

Figure 3.
Subdomain SD1—molten zone inside the head.

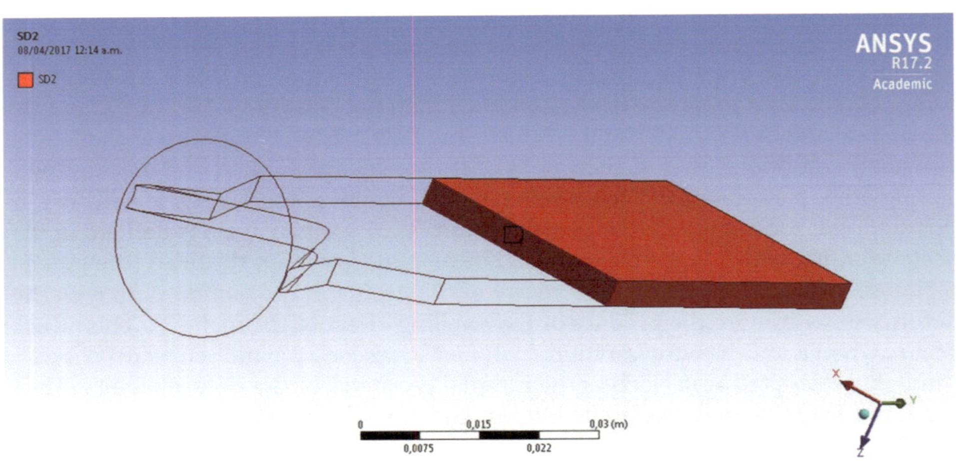

Figure 4.
Subdomain SD2—fade zone at the head output.

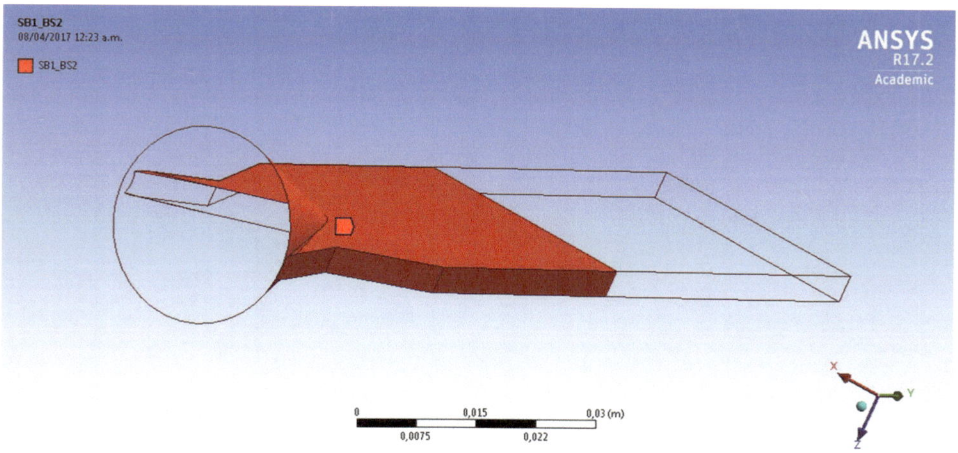

Figure 5.
SD1_BS2: outline of SD1.

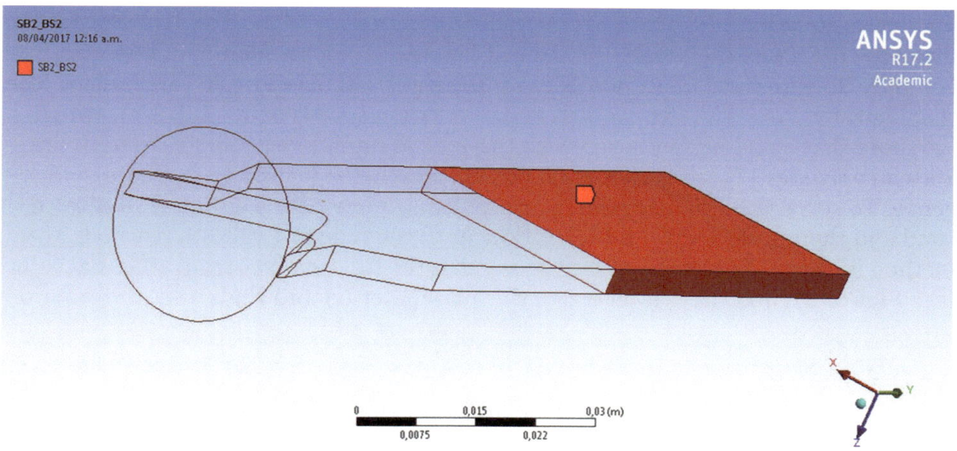

Figure 6.
SD2_BS2: outline of SD2.

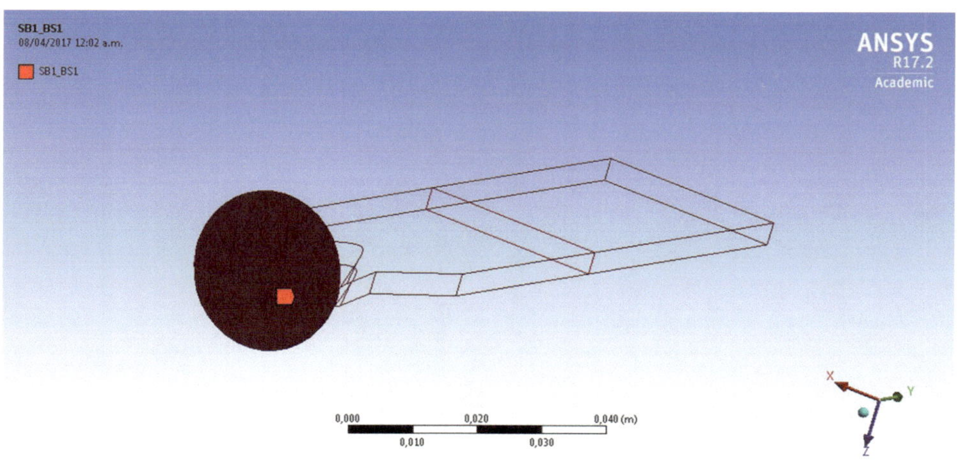

Figure 7.
SD1_BS1: head flow inlet surface.

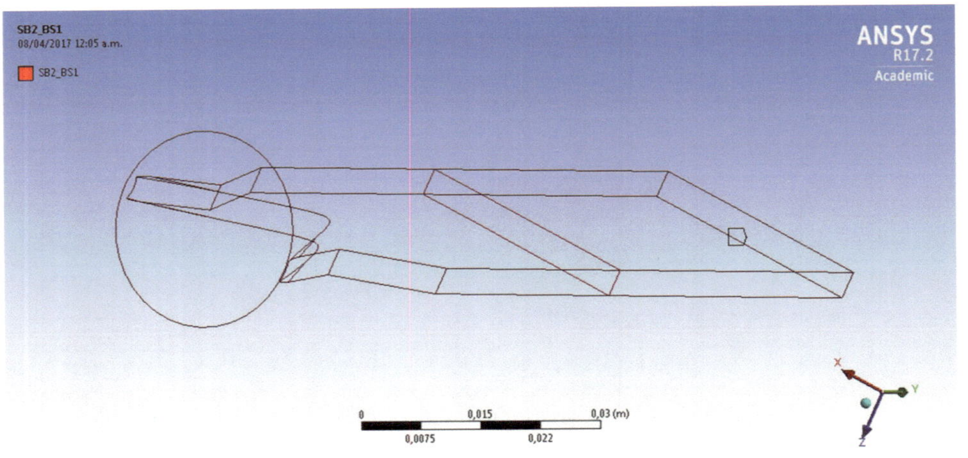

Figure 8.
SD2_BS1: head flow outlet surface.

kinematic equation f (s) that introduces nonlinear terms in the problem that lead to the convergence of the model. As in our case, we have a problem of a numerically complex flow because it is a non-Newtonian fluid with low Power Law indices (the rheological values were determined at deformation speeds of 50–100 1/s), which are those that could be measured with the special capillary rheometer used in line with an extruder [7], an incremental numerical scheme is used to facilitate convergence. To solve these highly nonlinear problems, a low flow solution is first calculated and then projected for a higher flow until the required value is reached. This method is called "Evolution" that is to say that the flow evolves from an initial value (see **Figure 9**), and then, it increases this parameter (*s*) and finds a second solution

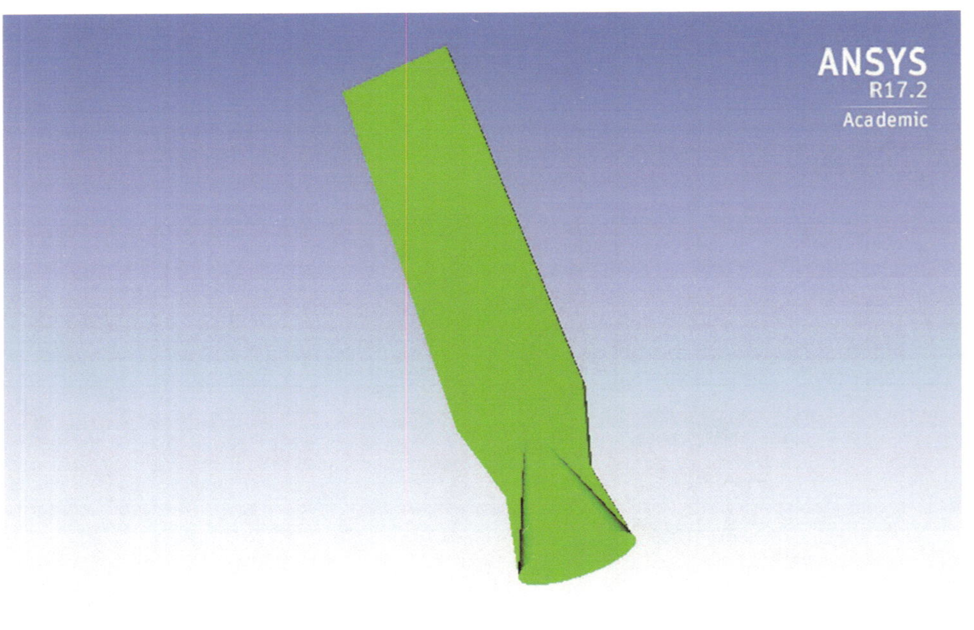

Figure 9.
Evolution of the S parameter.

and so on until the model is optimized, and there may be several intermediate steps and in each of them, the solution of the previous step is obtained and so on repeatedly until convergence is achieved. The concept can be applied to different boundary conditions (flow, temperature, drag force, amount of mass slip, etc.) and material properties (shear thinning index, relaxation time, specific heat, etc.). This process is carried out automatically and the increments are adapted in the same way, until the best solution is found. $Q = Qnom.f(s)$, $Qnom$ is the value of nominal, flow $f(s)$ is the evolution function and "S" is the evolution variable.

Next, in solution, the computational run is made to calculate the results for the model defined in the previous steps.

Finally, the post-processing is carried out with which the results of the calculations performed are obtained, such as the values of the distribution of speeds, shear deformations, pressures, contours of the profile at the exit of the head, and the graphs of each of them for analysis.

2.3.3 Experimental design and construction of the calibrator

The caliper is experimentally designed and drawn with 3D solid computer-aided design software (**Figure 10**).

The different parts that make up the calibrator are built and assembled (**Figure 11**), placing the assembly in the chamber or calibration-cooling pan and leaving the fixing screws of the upper part unadjusted, just positioning them, leaving the necessary play to enter the profile for calibration.

The gauge defines the final external dimensions of the profile [1], so that the profile in the molten state at the outlet of the head, at temperatures close to 200°C,

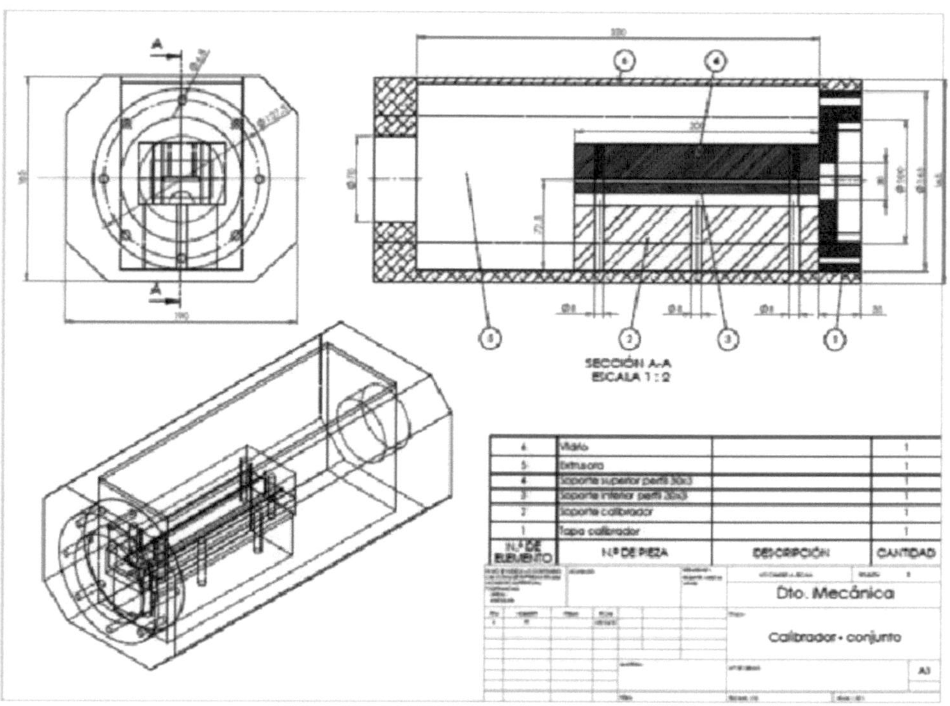

Figure 10.
CAD-3D caliper drawing.

Figure 11.
Assembled calibrator.

enters the gauge, where vacuum is applied and the material copies the internal dimension calibrator.

On the other hand, when the calibrator is immersed in a pan with water at 20°C, the profile cools, solidifying.

In order to define the internal dimensions of the caliper, it was taken into account that the molten material, when it cools, contracts and in the case of rigid PVC compounds it can reach high values of up to 4%.

In addition, another decrease in the cross-sectional area must be taken into account due to the stretching caused to the material by the dragging carried out by

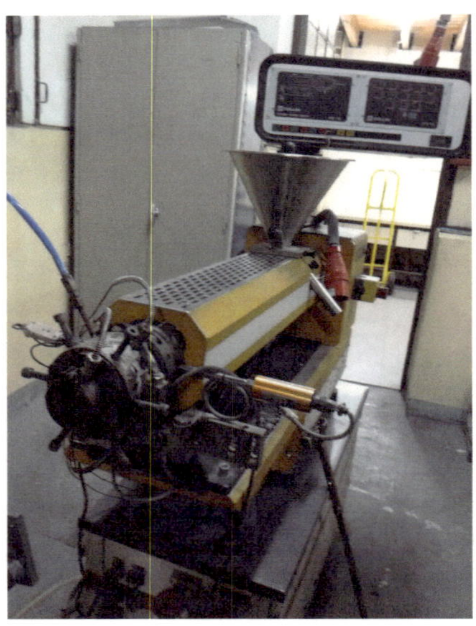

Figure 12.
Collins extruder.

the draft equipment, which can be between 5 and 10% for low-thickness rigid PVC profiles [8].

2.3.4 Extrusion tests

Extrusion tests are carried out using a Collins single-screw extruder (**Figure 12**) with a 2:1 compression ratio screw, the auxiliary equipment of the same brand, such as the vacuum and cooling calibration trays (**Figure 13**), and the system drag or pull (**Figure 14**).

Figure 13.
Calibration and cooling pan.

Figure 14.
Drive unit.

3. Results

3.1 Rheological values

The rheological values were obtained from the test carried out by technical personnel of the company ALFAVINIL S.A., manufacturers of PVC compounds, using a RheoDrive 7 capillary rheometer, model: Rheomex 19/25 OS, in line with an extruder, whose screw has a ratio of compression 2:1, L/D: 25 and a capillary matrix of slit H: 2.0 mm, W: 20 mm and with a profile of temperatures of the material of 150, 170, and 190°C in the extruder and 200°C in the head.

3.2 Simulation

3.2.1 Speed distribution in different sections

In the post-processing stage of the simulation, cross sections are defined at certain distances in the flow channel, to analyze the distribution of the velocity vectors in the same section and in each of them (**Figure 15**).

It can be seen that in the central regions of the sections in the feeding zone (conical zone), these velocity vectors have a higher value (celestial zone) than in the contours since, as the flow of the material is close to the inner wall of the flow channel, the velocity is zero (blue zone), this is consistent with what happens in reality when a fluid circulates through a conduit, and the velocity profile will be maximum in the center and tending to zero in regions close to the inside walls of the same.

As the flow of the molten material progresses, velocity vectors can be observed in the central areas of each cross section of greater magnitude (more intense green color), and this is due to the abrupt change of the geometry in the flow channel going from a conical rectangular; in particular, if we analyze the fourth cross section

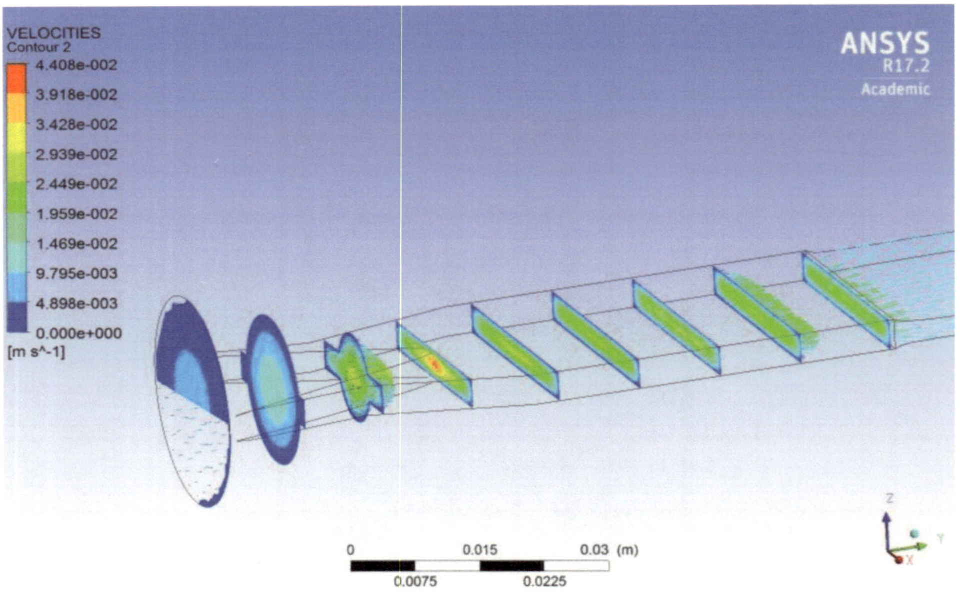

Figure 15.
Distribution of speeds inside and at the outlet of the head.

from the entrance of the flow to the head, it is observed that in the center there is a zone of velocity vectors of even greater magnitudes (red color).

Variations in the velocity vectors at each point of the same section cause the flow channel not to comply with the design principle of "minimum volume" of the heads; that is, the melt must reach all the points at the same time of the cross section or with the same speed.

These differences between the sliding of the fluid in the central zones with respect to those that are closer to the inner wall of the head cause internal shear stresses in the molten material that could later lead to the appearance of failures in service; this becomes more critical in profiles with complex shapes and in the particular case that between width and thickness, there is an important dimensional difference.

3.2.2 Distribution of shear deformations

In the graph shown below (**Figure 16**), it is observed that in the area where the geometry changes, shear deformations are generated (all colors are very high except blue), and this indicates that in these places, the molten material can get to degrade, especially in our case in which rigid PVC is used, due to the generation of high localized temperatures due to friction and shear.

In order to reduce or prevent this detrimental effect from being generated, it is advisable to modify the geometry of the head flow channel so that the transition between the two zones is smooth.

The simulation results predict in this case the detrimental effects that occur with the current head design, so that if it is necessary to build a new head, the geometry must be modified, so that the shear deformations are low and if possible, their

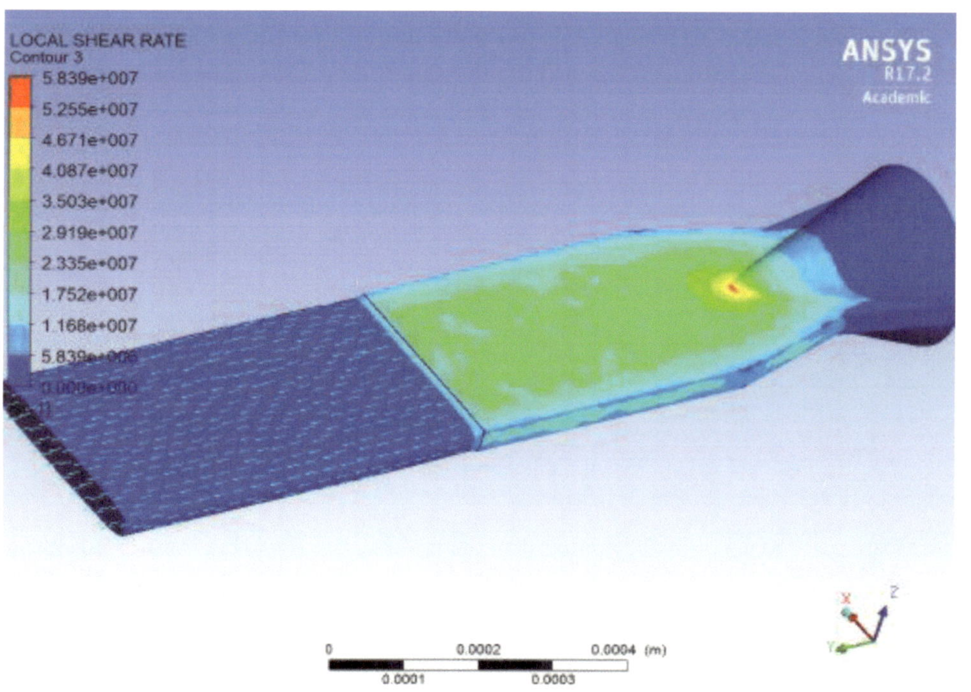

Figure 16.
Distribution of shear deformations.

values are between 50 and 100 1/s, which is advisable to process this type of material and as described in the rheological test, they are the values where the Power or Ostwald Law is valid.

Another analysis that is carried out is on the contour areas at the entrance and exit of the molten material from the head, in which the shear deformation is zero, especially if a final parallel area of the matrix is left sufficient to have a swelling of the die, minimum fade, and within the recommended parameters [6].

3.2.3 Comparison of original profile contours vs. cast profile outlet head

Another important result of this simulation is obtained: the contour of the profile in the molten state at the outlet of the head at a distance of 60 mm.

To do this, the orientation of the axes of the channel geometry is modified in the post-processing stage, eliminating all contours and surfaces except for the contours of the output profile of the head (original) and the profile of the molten material at the output, from the head at a distance of 60 mm.

By superimposing both contours, the geometric and dimensional differences that exist between them can be seen and therefore, this image is exported and said differences are dimensioned (**Figure 17**).

They are due to the aforementioned incidence of the swelling effect of the melt that causes an increase in the cross-sectional area of the flow of the melt at the outlet of the head.

In a previous simulation, a channel geometry model was used with an outer zone at the head outlet of 10 mm, with which the deformations were greater, there were important geometric and dimensional differences, and the profile could not have been calibrated or manufactured, because it would have stuck in the caliper, thus optimizing the design of the model by increasing the length of the area at the exit of the cast profile with a new simulation at a distance of 60 mm.

The dimensional variation at a distance of 60 mm from the head can be seen in **Table 3**.

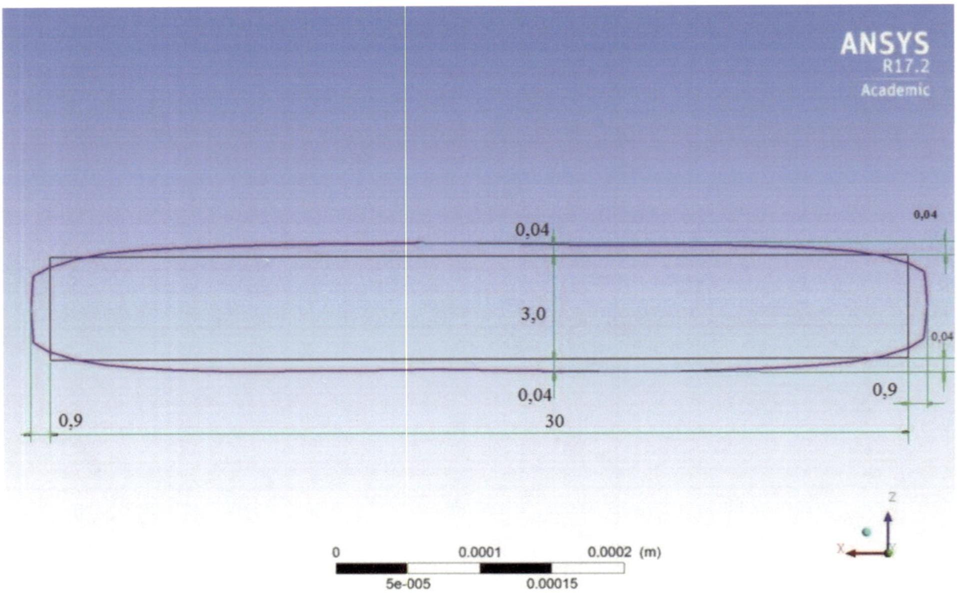

Figure 17.
Comparison of original profile contours (black) and head outlet (blue).

Characteristics	Original profile	Profile molten at heat out	
	mm	mm	Dimensional differences (%)
Thickness	3.0	3.08	2.7
Wide	30.0	31.8	6.0

Table 3.
Dimensional variations between the original profile and the head outlet profile at a distance of 60 mm.

Therefore, maximum deformation values were achieved due to the swelling effect of the melt within the recommended parameters, which are 3–6% for thicknesses of rigid PVC profiles between 3 and 4 mm thick [6].

3.2.4 Pressure distribution

The graph with the pressure distribution was obtained, since at the entrance of the flow of the molten material to the head the pressure is maximum and is the one generated in the extruder at the exit of the dosing zone, with values that depend on flow rate and material temperature, as well as screw and die design (back pressure).

The pressure decreases as the flow advances inside the head, as can be seen in **Figure 18**, so that at the exit of the matrix (blue area), it is zero, as it should be since the flow of the material cast is stress free.

It is observed that the pressure value obtained from the simulation in the area where the pressure sensor is located in the head (light blue color) has values between 142 and 68 bar (in **Figure 19** the pressures are expressed in Pascal (Pa) and

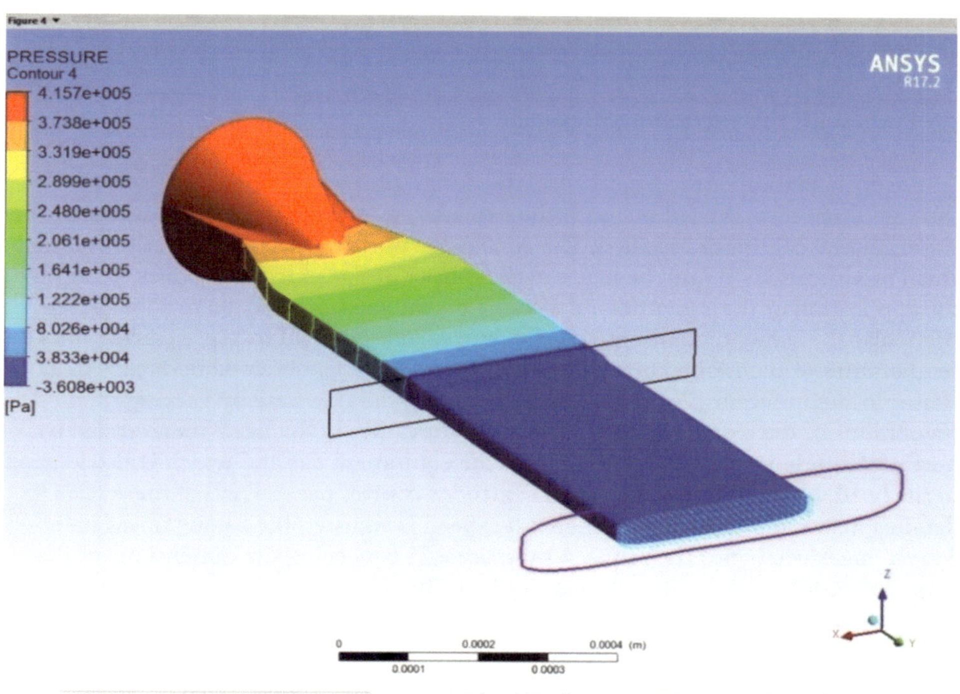

Figure 18.
Pressure distribution inside and outside the head.

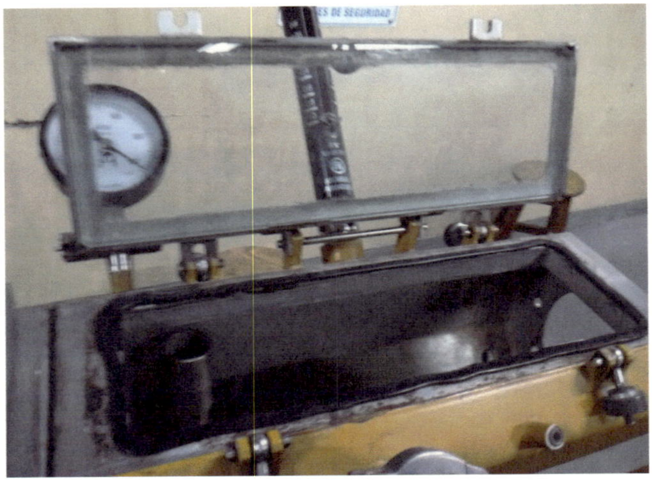

Figure 19.
Cooling and vacuum calibration pan.

the The value measured with said pressure sensor is 51 bar for a flow rate of 8.5 kg/h and a temperature measured with another sensor in the head of 200°C.

This difference between the simulated and real pressures of the process may be due to the high shear deformations that occur inside the head and that when using the Power Law or Ostwald model, the same is for deformations, much lower and on the order of 50–100 1/s.

The large difference between the shear deformations obtained as a result of the simulation (between 5.2107 and 5.6107 1/s in the highest values of red color) and that used by the Power Law affects the results of the pressure distribution that delivers the software.

3.3 Tests with the extruder equipment

An optimization of the process and the design of the calibrator is carried out through three tests carried out with the extruder, its auxiliary equipment, and the experimental calibrator installed. The conclusion of these machine tests showed that small modifications should be made in the caliper design, such as grooves to improve the application of the vacuum, and also the machine parameters were adjusted, in particular the screw revolutions and the drag speed. The said tests are carried out at a temperature of the molten material in the head of 200°C, obtained by means of a sensor located therein (**Figure 12**) and modifying the flow rate by varying the revolutions of the extruder screw. The pressure values in the head are recorded by means of another sensor (**Figure 13**) that the equipment has and which is also located in the head, observing its value on the extruder control panel. Each of these tests is detailed below: 3.3.1. First test: The screw speed is adjusted for a mass flow rate of 5 kg/h, measuring with the sensor a pressure = 55 bar, being the dimensions of the extruded profile at the exit of the gauge of width = 26 mm and thicknesses: 2.6 mm on one side and 2.9 mm on the other side of the width. It is determined that the cause of not reaching the required dimensions (width = 30 mm and thickness = 3 mm) is a vacuum deficit in the calibration pan by vacuum and cooling (**Figure 14**). The vacuum is generated by a pump located in the water tank that is located at the bottom of the vacuum and cooling calibration pan and this vacuum is measured with a pressure measuring instrument or manometer (**Figure 19**).

3.3.1 Second test

The screw revolutions were adjusted to 48 rpm so as to obtain a mass flow rate of 5.7 kg/h and a pressure of 47 bar was measured.

To increase the vacuum, a calibrator will be made with internal grooves that communicate the place (hole) where the final geometry of the profile is defined, with the room where the mentioned vacuum is applied.

The dimensions of the profile were measured at the exit of the caliper and the following values were obtained: width = 29.6 mm and thicknesses in the central area of the profile = 3 mm and at each end of the width = 2.7 mm.

Therefore, although the dimensional value of the width is improved, the geometry is still not copied well on the sides and also the geometry is not rectangular, since the thickness in the central part is 3 mm and decreases toward the edges reaching a value at both ends of 2.7 mm; therefore, the process is not yet suitable to correct this defect.

3.3.2 Third test

The screw revolutions are again modified to 65 rpm, obtaining a mass flow = 8.5 kg/h, and it is noteworthy that this flow value is the same that is used as one of the input variables in the simulation software.

The measured pressure was 51 bar (in this case, there is a difference with the values obtained in the simulation, since for the area where the pressure sensor is located, the results were between 142 and 68 bar.

The increase in the mass flow, accompanied by an increase in the drag speed and using the calibrator with the grooves to improve the application of vacuum on the flow of molten material that comes from the head and that is calibrating and cooling, achieves the desired geometry of the outlines, which is the output of the caliper.

On the other hand, the required dimensions are achieved with an error of less than 5% at the ends of the width, since it was measured with a coliza foot that the width = 30 mm and at the ends = 2.9 mm, that is, the error was 3.3%.

The flow measurement was performed by taking the amount of material that was extruded per unit of time and it coincides with the flow rate that was used as an input variable in the simulation software, that is, 8.5 kg/h, working with the screw at 65 rpm.

The sensed pressure was 51 bar and that obtained from the simulation software ranged between 142 and 68 bar, and the explanation of these differences is due to the fact that the Power Law model used is for low deformation speeds and in the simulation, said deformation shear was very high, especially in the area where the geometry of the channel was modified from conical to rectangular (transition zone).

4. Discussion

The result of the first simulation using a geometry of the flow channel with an area at the outlet of the head of 10 mm (**Figure 20** area in green color) generated high deformations due to the swelling effect of the melt, which means that there was no possiblity to calibrate and manufacture the profile to the required dimensions, as explained above.

Therefore, the model is modified by increasing the length of said area to 60 mm and it is simulated again, obtaining deformation values that are within the

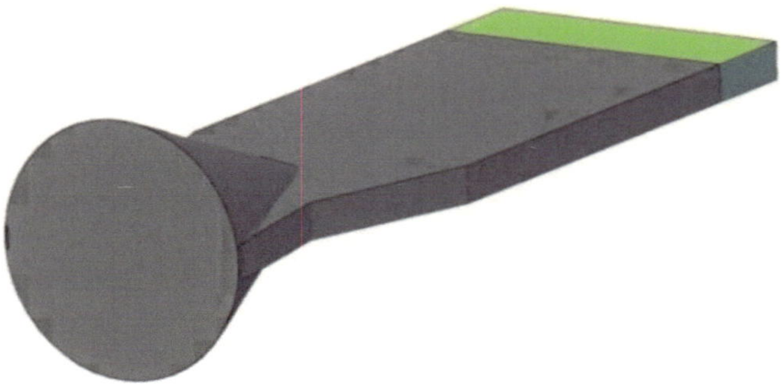

Figure 20.
10-mm outlet area flow channel geometry (green area).

recommended range. This length of the area of the molten profile outside the head is the one that is recommended to use for the extrusion of the same; that is, the distance between the head and the gauge should be of that magnitude to ensure proper calibration and not having clogging when the profile in the molten state enters the calibrator. When a vacuum is applied in the calibration-cooling pan, the molten material in the calibrator increases its cross section by copying the internal dimensions of the same and advancing with a continuous flow by means of the drag force, while it solidifies when immersed in water at 20°C. In addition, the software uses overlock optimization in the said exit zone, which is the one shown in this work, since by increasing the length of the free-flow zone at the exit of the head, the deformation due to the swelling effect of the fade decreases.

5. Conclusions

Through simulation, a 60-mm-long profile geometry model is obtained at the outlet of the head, thus minimizing the swelling of the melt and obtaining a deformation with dimensional differences that are within the recommended percentages, as described in Section 3.2.3.

The 60-mm length obtained from the simulation is recommended to use in the extrusion process to overlap the distance that must exist between the head and the caliper in order to avoid problems of clogging of the molten material that enters the caliper and apply vacuum.

From the results of the simulation, it is determined that the distribution of the velocity vectors in each of the cross sections is not homogeneous and also the shear deformations had very high values, which cause internal stresses to be generated in the profile that causes their failure in service.

Therefore, the geometry of the head flow channel should be redesigned to improve these parameters that influence the final properties of the profile, such as the generation of high internal stresses that affect the quality of the product.

An optimal parallel zone can also be defined, because its length influences the swelling effect of the melt; that is to say, by increasing it, a greater relaxation of internal stresses is obtained and consequently a less deformation of the material at the exit of the caliper [7]. It was possible to carry out the rheological test and obtain the k and n values of the Power Law for the rigid PVC compound of the present work using a special capillary rheometer connected in line with an extruder and they could be used in a simulation software that uses fluid dynamics. Computational and the direct extrusion method [9] is used to optimize the design of a head of a rectangular profile in rigid PVC.

Therefore, the predictive capacity of the ANSYS Polyflow simulation software is verified for this work and with it, it can be applied for the simulation of more complex profile geometries.

Finally, a specific calibrator is designed and built for this profile and in only three tests with the extruder equipment, the operation of the same and of the process is optimized, verifying that the flow used in the simulation is correct, obtaining a profile with the geometry and final dimensions required and with a dimensional error of less than 5%.

Acknowledgements

Special thanks to the University Technician Gustavo Pierson, alIng. Federico Tirapelli, and Technician Javier Ríos Zabala, staff of the Higher Polytechnic Institute-U.N.R. for their collaboration in the design, construction of the gauge, and participation in the extrusion process.

To Dr. Patricia Frontini from INTEMA, UndMP and Dr. Javier Signorelli from IFIR, CONICET Rosario for their contributions and suggestions.

Finally, to Dr. Fernando Angiolini and Mr. Leonardo Tort, technical and managerial staff of the company ALFAVINIL S.A., who selflessly lent their collaboration to carry out the rheological tests and provided the material to be able to carry out the extrusion tests.

Author details

Carlos José Salvador Tomassini
Department of Plastics and Elastomers, University of Rosario, Rosario, Argentina

*Address all correspondence to: carlos_tomassini@hotmail.com

References

[1] Szarvasy L et al. Computer aided optimization of profile extrusion dies. International Polymer Processing, the journal of the Polymer Processing Society. 2000;**15**(1):28-39

[2] Srinivasa R et al. Extrusion simulation and experimental validation to optimize precision die design. In: ANTEC 2004. IL: Northern Illinois University; 2004

[3] Vlachopoulos J, Fattamann G. The Role of Rheology in Polymer Extrusion. Hamilton, ON, Canada: Department of Chemical Engineering, McMaster University, Polydynamics, Inc.; 2015

[4] Titow MV. PVC Technology, 4th ed. England: Elsevier Applied Science Publishers; 2012

[5] Zhao W. Finite element simulation of 3D unsteady viscoelastic free surface flow with level set method. In: ANTEC, Society of Plastics Engineers Annual. 2010

[6] Michaeli W. Extrusion Dies for Plastic and Rubber. 3rd ed. Munich, Germany: Hanser Publishers; 2003

[7] Goncalves ND et al. Design of complex profile extrusion dies through numerical modeling. Journal of Non-Newtonian Fluid Mechanics. 2013;**200**: 103-110

[8] Limper A, Fattmann G. Application of an Online Rheometer to Evaluate the Melt Properties of PVC. Germany: Institut für Kunststofftechnik, University of Paderborn;

[9] ANSYS. ANSYS Polyflow Tutorial Guide. Release 17.0. ANSYS, Inc.; 2016

Section 2

Carbon Fiber Reinforced

Chapter 3

Unidirectional Carbon Fiber Reinforced Thermoplastic Tape in Automated Tape Placement Process

Svetlana Risteska

Abstract

Thermoplastic matrix composites are finding new applications in the different industrial areas, thanks to their intrinsic advantages related to environmental compatibility and process-ability. The tape placement process is one of the few techniques that have the potential to continuously process thermoplastic composites in large industrial applications. Fiber-reinforced thermoplastic tapes are subjected to high heating and cooling rates during the tape placement process. The application of laser heating for the tape placement process requires a thorough understanding of the factors involved in the process. Qualitative experimental analysis is presented to identify the important phenomena during the tape placement of carbon (PEEK, PEKK, PAEK PPS) tapes. The present chapter focuses on the input parameters in the process of manufacturing composite parts. The mechanical performance of the final parts depend on a number of parameters. It should be void-free and well consolidated for reliable use in the structure. In the present work, it is becoming increasingly wiser to introduce the production of high-quality laminates, using laser AFP and ATL with quality consolidation during the laying process. The experimental results in this chapter help to better understand the consolidation process during LATP.

Keywords: reinforced, fiber, thermoplastic, laser-assisted automated tape placement (LATP), in situ

1. Introduction

The thermoplastic materials do offer a great advantage over thermoset prepreg and dry fiber materials [1, 2]. For this advantage to be realized, thermoplastics must better compete with thermoset prepreg production for large structural parts in aviation. Today, most thermoplastic research has centered around full *in situ* consolidation to eliminate post-processing for curing. The main restriction of full *in situ* consolidation is on the speed (and productivity) of the LATP/LAFP process, which is limited by the physics or chemistry of thermoplastic materials. This method places all the demands of final part quality (low porosity, proper crystallinity, and autohesion) on the AFP process.

Robotic automated fiber/tape laying processes are used to produce high-performance composite components from unidirectional precast materials. In this procedure, a robot is most often used to lay the lanes along pre-defined paths, which provides a high degree of freedom in designing the final product. The process of automated laying of strips consists of automated laying, that is, stacking of layers on top of each other on a flat or complex-shaped tool, which is called a mandrel. Pre-layering is associated with the application of heat and pressure. It is important during this procedure to allow the strips to consolidate directly during installation, which will prevent additional time and energy from being spent on post-consolidation [3–7].

Most of the research for automated production is oriented toward the application of thermoplastic prepress materials, although today many responsible and load-bearing composite parts obtained with thermoset materials are in use [5]. The reason for such research is that when processing a thermoplastic prepreg, the material can be bonded on-site directly during layup (placement). However, when processing a thermoplastic prepreg, just like when processing a thermosetting prepreg, many challenging problems occur, such as gaps and overlaps between the tapes, which affects the final characteristics of the material, which requires in-depth research. The research done for the application of robotic processes for automatic laying of fibers/tapes is expected to provide better obtaining of high-quality structures from composite materials with a new process called on-site consolidation (*in situ*) [6–9].

The advantages and benefits of the application of these procedures for automated laying of fibers/tape for obtaining parts of composite materials are the following [4]:

- Higher productivity in production,
- Better quality of the composite,
- Less unused waste material,
- Less labor,
- Shorter production time,
- Reduced qualifications of the professional production staff,
- Accuracy and repeatability of the product,
- Software solutions.

The concept of *in situ* consolidation is simple. The inlet thermoplastic pretape in contact with the substrate or mandrel under sufficient pressure and temperature above the melting point is consolidated and crystallized upon cooling at the controlled or uncontrolled speed [10–25].

2. Manufacture of the final product with LATP/LAFP technology

The steps for the production of a quality final product in this process are given in **Figure 1**; in each subchapter, all the steps are elaborated in more detail.

2.1 Selection of raw material UD thermoplastic prepreg

A very important process parameter is the material used in LATP/LAFP processes. The production process for obtaining composites from thermoplastic materials mainly depends on the viscosity of the resin and therefore, impregnation is very

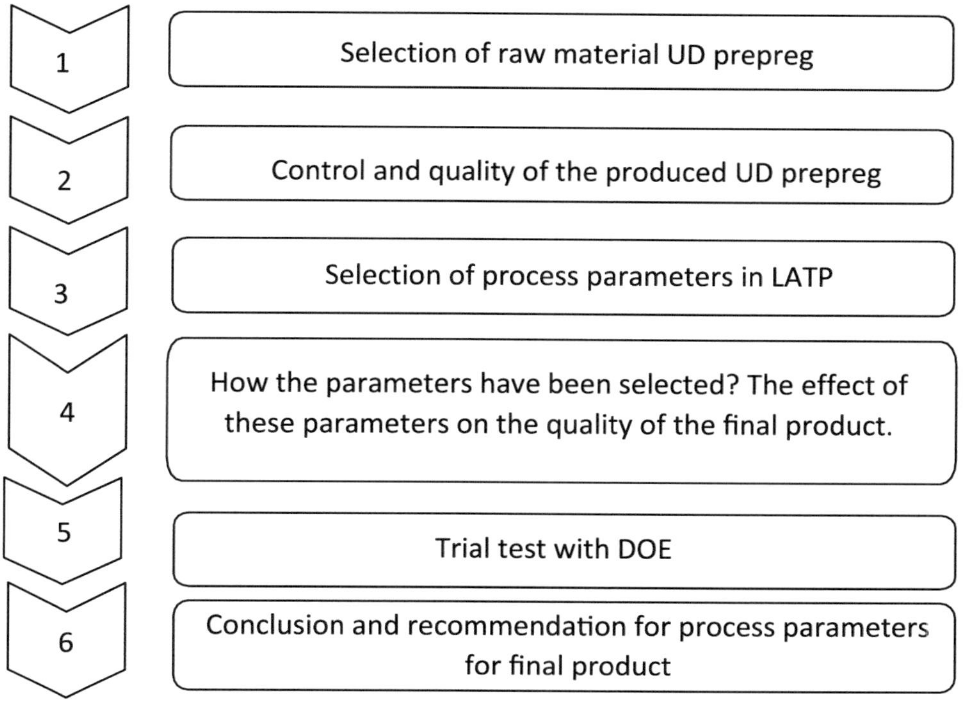

Figure 1.
Procedure for making a quality product with LATP technology.

important in the processing of thermoplastic materials, that is, the characteristics of the raw material. The characteristics of the thermoplastic prepreg determined by the content of pores, volume fraction of fibers, degree of wetting of the fibers, uniformity, surface roughness, etc. determine the potential of the process. In order to make a suitable and good quality product from thermoplastic prepreg, the first thing that needs to be done is to look for suitable raw material. Thermoplastic tapes, for the purposes of this LATP/LAFP process, consist of unidirectionally aligned carbon fiber tows in widths of up to 12 inches/305 mm, prepregged with a thermoplastic resin. The resins most commonly used in aerospace and other high-performance applications are the following high-performance thermoplastics—polyether ketone ketone (PEKK), polyether ether ketone (PEEK), polyaryl ether ketone (PAEK), polyphenylene sulfide (PPS), and polyetherimide (PEI). Some manufacturers offer tapes prepregged with commodity thermoplastic resins, such as polyamide (PA6), polyether sulfone (*PES* or PESU), polypropylene (PP), and others, but these are generally considered unsuitable for large aerostructures.

Thermoset tapes are available in widths of up to 60 inches/1524 mm and can go thousands of meters without a defect, while thermoplastic tapes typically top out at 12 inches/305 mm and show as many as 30 defects in just 210 m ft. It is because the thermoplastic UD prepregs rely on a powder-based application process that is more difficult to control and can create resin-rich and dry areas. Such nonuniformity can lead to problematic interplay porosity.

Some materials used in this technology from different manufacturers are given in the **Table 1** and **Figure 2**.

The key point is that high-quality tapes, with low void content, are an enabler of fast, automated processing of high-quality composites. Tapes with high levels of voids will require longer consolidation cycles to produce high-quality parts. Today there are several manufacturers of thermoplastic UD tapes, such as Suprem,

Matrix	Supplier	Type	PAW (g/m^2)	Carbon fiber	Thickness (mm)
PPS	Ten Cate	UD TC1100 PPS/AS4UD	220	Standard	0.14
	Suprem	Suprem T 60% AS4 / PPS-214	220	Standard	0.14
	Barrday	C/PEEK TU200–145-HM63–37-12	220	IM	0.14
	Barrday	TU0110–60% PPS DIC MA520/AS-4D	220	Standard	0.14
PEEK	Ten Cate	TU0200 PEEK VICTREX 150-AS4D	220	Standard	0.14
	Ten Cate	TU0200 PEEK /H800	220	IM	0.14
	Ten Cate	TU0200 PEEK /IM7	220	IM	0.14
	Suprem	Suprem T 60% AS4 / PEEK-150	220	Standard	0.14
	Suprem	Suprem T 60% AS4 / PEEK-151	220	IM7	0.14
	Barrday	TU0200-PEEK VICTREX 150-AS4D	220	Standard	0.14
	Barrday	TU0200-PEEK VICTREX 150-IM7	220	IM	0.14
	Barrday	MC0000-KG is C/PPS TU0110–145-HM63–37-12	220	IM	0.14
	Tejlin (T.Tenax)	TPUD PEEK-HTS45	220	Standard	0.14
	Tejlin (T.Tenax)	TPUD PEEK-6-34-IMS65 P12 24 K-UD	220	IM	0.14
	Solvay	Solvay's APC 2 (PEEK-FC)	220	IM7	0.14
	Solvay	Solvay's APC 2 (PEEK-FC)	220	Standard	0.14
PEKK	Ten Cate	Ten Cate'UD TC1320 PEKK/ AS4	220	Standard	0.14
	Suprem	Suprem T 60% AS4 / PEKK-7002	220	Standard	0.14
	Suprem	Suprem T 60% IM7/ PEKK-7003	220	IM	0.14
	Suprem	Suprem T 60% AS4 / PEKK-7002	300	Standard	0.19
	Barrday	TU300 AS-4D PEKK ARKEMA 7003	220	Standard	0.14
	Solvay	Solvay's APC (PEKK-FC)	220	Standard	0.14
PAEK	Ten Cate	TC1225 Cetex® TC1225 LM PAEK, AS4D	220	Standard	0.14
	Tejlin (T.Tenax)	PAEK/CF Teijin-UD	220	Standard	0.14

Table 1.
UD tape thermoplastic prepreg from different suppliers for LATP/LAFP.

Barrday, Tejlin (Toho Tenax), Toray advanced composite (Tan Cate), Solvay, Celstrane, etc. They all offer different thermoplastic tapes as mentioned above with different matrices and of course with different carbon fibers (**Figure 2**). So, before

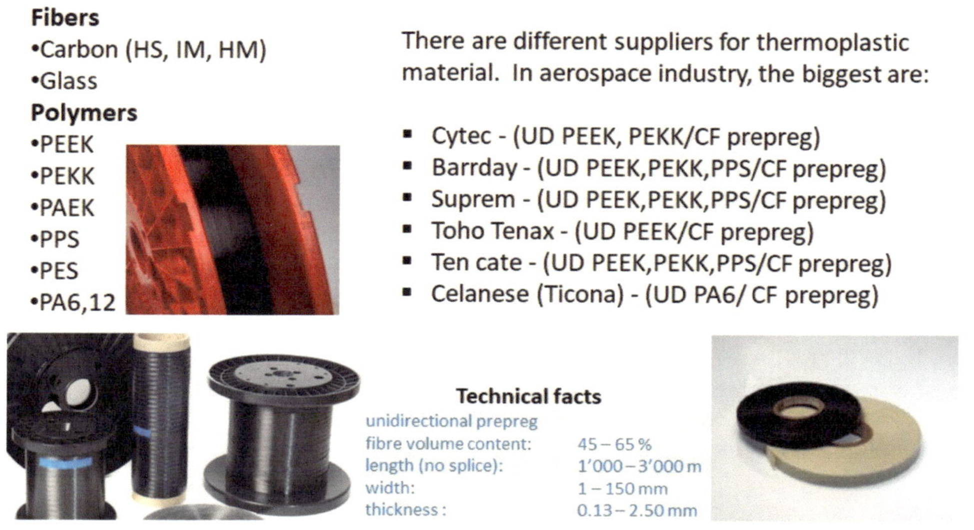

Figure 2.
Materials for LATP/LAFP technology suppliers for UD thermoplastic prepreg.

we start making the part, we need to know what type of fiber we need and which matrix (depending on which characteristics the final product we want to get). After determining the type of fibers and the matrix, the next step is to procure materials from several manufacturers. After that, the second phase of quality control of the raw material begins.

2.2 Control and quality of the produced UD prepreg

For the *in situ* consolidated LATP/LAFP process first main is the quality of the raw material, which can vary greatly from supplier to supplier and for each material matrix. As mentioned earlier in the production of thermoplastic tape can often occur defects of the tape itself (especially porosity) which requires a different temperature, greater consolidation, and melting time during the LATP/LAFP process. Therefore, the raw material must be tested. What is meant by quality control of raw material in this process? It is an examination of purchased raw material:

a. Inspection of the raw material quality documents supplied by the material supplier.

b. Tests and control of the purchased raw material:

- degree of crystallinity of raw material,
- void content of raw material,
- content of resin/fiber of raw material,
- surface roughness of raw material,
- microphotography (surface and cross-section) of raw material.

c. Controls of the characteristic of purchased UD thermoplastic prepregs.

d. Preparing the report of the raw material examination.

e. Decision of raw material acceptance/rejection.

The thermoplastic tapes are typically not as consistent as thermoset materials, the uneven resin application can produce regions of resin richness, which can be both helpful and detrimental. In order to see a homogeneous distribution of the resin in the raw material, microphotographs of the procured raw materials are made with the help of an optical microscope (**Figure 3**).

The following figure shows some optical images from more raw materials UD thermoplastic prepreg:

In LATP/LAFP process, the **surface roughness** of the raw material has the biggest impact. Complete intimate contact is achieved if all the unevenness is the same

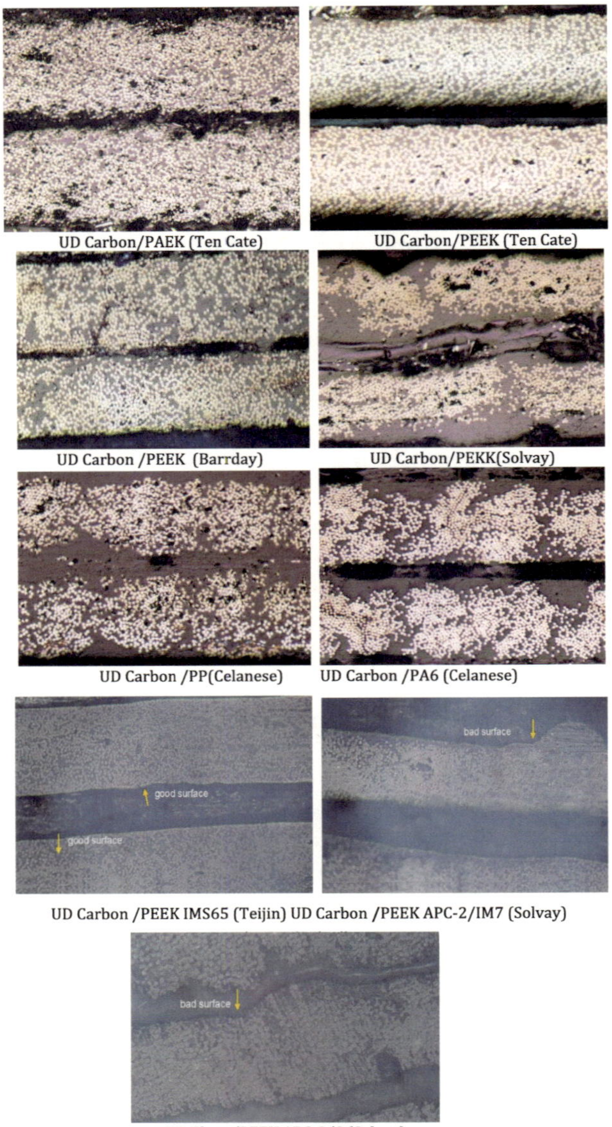

Figure 3.
Microscope images of some raw material (UD thermoplastic prepreg).

size, that is if all the unevenness is pressed to the same order [19]. The **presence of voids** in the starting material requires a longer heating time and greater pressure during the process of laying a thermoplastic strip with a laser. It is, therefore, necessary to know the percentage of voids on the tape that will be used in the process.

The **resin content** in the thermoplastic band has a great influence on the mechanical properties of the finish, but it also has an effect on the percentage of crystallinity and the heat balance for the LAFP/LATP process.

Determining the crystallinity of the raw material for the LATP / LAFP process is important to determine the temperature of the tool and to compare it with the crystallinity of the final product.

2.3 Selection of process parameters in LATP/LAFP

In general, the LATP/LAFP process using laser heating involves many parameters that affect the results of the process. To obtain quality parts, the combined effect of the main process parameters must be studied and analyzed. **Table 2** and **Figure 4** show the factors that affect the output of the automated fiber/tape laying process by using a laser heater. In the same table is entered a part where it is said which factor how much they influenced the final product and at what.

The association of semicrystalline thermoplastic tapes with the LATP/LAFP process has been analyzed in many literatures [6, 7]. Many studies have been done to determine the temperature distribution near the connection zone in order to determine the optimal angle of incidence of the laser beam to achieve a good interface between the layers [10–38]. Recent studies show that automated thermoplastic laminating processes with a velocity even greater than 200 mm/s produce composite

1. Raw material (UD tape)	
1.1. Tape impregnation quality (X)	1.3.Tape Surface Roughness (R) ($)(@)
1.2. Tape thickness (ttape)	1.4.Tape Tension (τ)
2. Mandrel	
2.1. Mandrel Curvature (DM) (@)	2.3.Mandrel Thermal Conductivity (kM)
2.2. Mandrel Temperature (TM) (#)	
3. Compaction roller	
3.1. Compaction Force (F) (@)	3.4. Roller Diameter (Dr)
3.2. Compaction Area (S)	3.5. Roller Thermal Conductivity (kr)
3.3. Roller interface Temperature (Tr)	
4. Heat source (laser)	
4.1. Beam intensity (I) ($)	4.4. Beam angle (β) ($)
4.2. Beam profile (Y) ($)	4.5. Incoming tape angle (α) ($)
4.3. Beam size and shape (W) ($)	
5. Process parameters	
5.1. Heat Transfer Coefficient (h)	5.3. Substrate Temperature (Ts) (@)
5.2. Placement Rate (v) (@)(#)	5.4. Tape Temperature (Tt) (@)

(@) has a high impact on the final product for voids.
(#) has a high impact on the final product for the degree of crystallinity.
($) has a high impact for (Tt) and (Ts)

Table 2.
Parameters that affect the quality of the final product of the process.

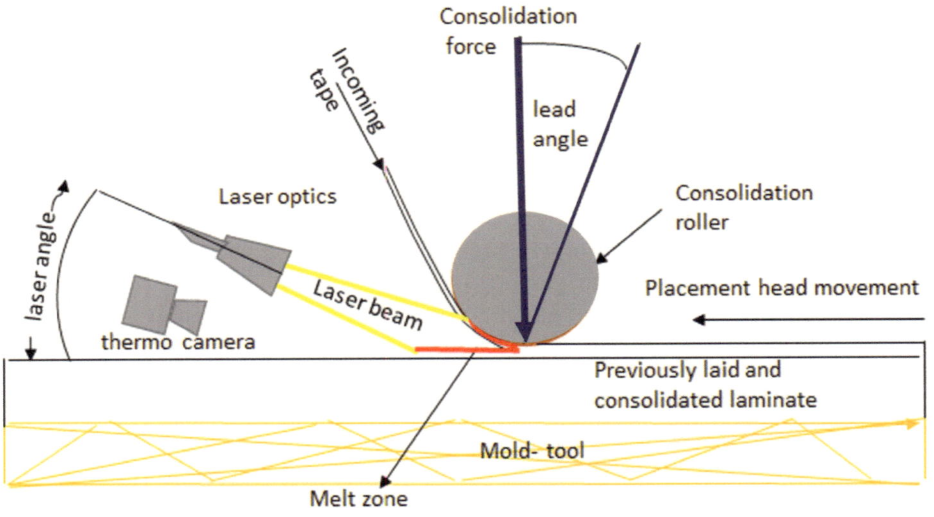

Figure 4.
Factors affecting the automated application process of laser heating.

flat slabs with properties close to those of the resulting flat slabs when consolidated in an autoclave [13–17].

2.4 How the parameters have been selected? The effect of these parameters on the quality of the final product

In research, some of these factors (**Table 2**) are taken as constants and some as variables. Most often, the selection of the variable parameters for the research within the tests and the analysis of their impact on the automated LATP/LAFP processes when laying the thermoplastic tape is given in **Figure 5**.

Raw material (UD tape) - The influence of the characteristics of the raw material was previously explained.

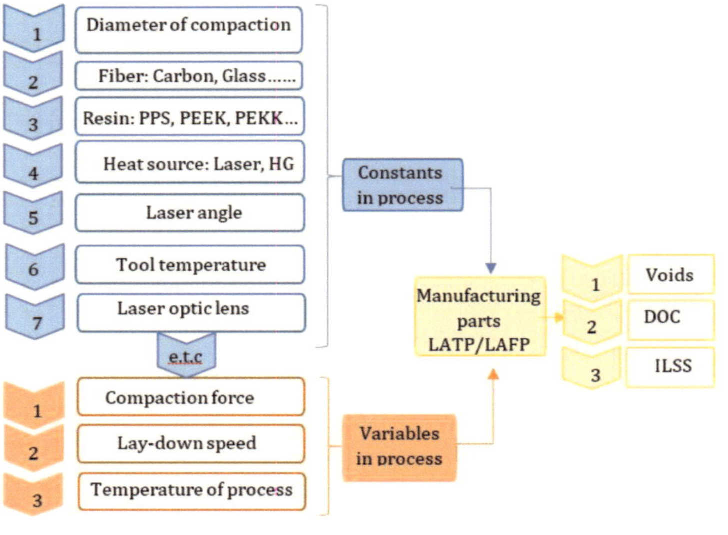

Figure 5.
Constants and variables for the LATP process.

Mandrel - In the process of passing this technology, important factors that affect the final product of the mandrel are as follows:

- Mandrel Curvature – Depending on the shape of the mandrel should
    - to determine the laying path - choice of lane width, for complicated and small curves the lane width should be smaller,
    - selection of the angle of the laser - depending on the shape of the mandrel, it is chosen how to position the laser so that we have even heating of the tape and the substrate
- **Mandrel temperature** – Numerous studies [20–26] have shown that heating the mandrel helps to release the stresses of the forces that occur in the cooling band so that there is no curvature of the flat parts. Mandrel heating is also important for the percentage of crystals in the final product. Experimental studies from many literatures have shown that the best mandrel temperature is if it is used 10–30°C above the glass transition temperature of the matrix used in the process.

Compact roller - The dimensions of this roller (diameter, width and thickness of the siliconized layer) are important because they affect the intimate contact between the layers in the final product.

The development of intimate contact between the layers, which is a prerequisite for connection, consists of equalizing the unevenness of the tape and the laminate. The initial surface unevenness is deformed under the action of heat and pressure. The time required to achieve intimate contact depends on the unevenness of the surfaces, the applied pressure, and the viscosity of the matrix which depends of course on the temperature. Due to the temperature dependence of the viscosity, the increase in temperature facilitates the development of contact. In regions where intimate contact is achieved, interdiffusion of polymer chains occurs due to accidental thermal movement. The interdiffusion process is generally explained by the mobility of the polymer chains. The polymer matrix consists of intertwined chains that have limited movement. Their mobility, and thus the degree of diffusion, increases with increasing temperature. In the case of semicrystalline polymers, the presence of crystalline regions can severely inhibit the interdiffusion process. Based on this, it can be noticed that the interface is one of the important parameters, and therefore the thermal aspects when laying the tapes are very important.

When processing thermoplastic tapes, the connection is the result of intimate contact and autohesion (direct-*bonding* or self-*bonding*). *Autohesion* begins after the onset of intimate contact. Different connection models have been proposed in the literature depending on the type of process [18, 30–35]. Some processes require a long processing time to complete the layers. However, for LATP/LAFP processes, the temperature of the polymer rises to the melting point before the start of intimate contact [35]. Stokes-Griffin and Compston [34] later found that the bonding of carbon fiber-based thermoplastics and PEEK using LATP/LAFP takes place at temperatures below the melting point and above the glassing temperature. They found that PEEK polymers are very amorphous during consolidation and require short processing times with LAFP or LATP processes using extreme cooling rates (1000°C/s), and the polymer is heated above the melting point before starting intimate contact.

Heat source (laser)

Today, industries and research institutions do a lot of research on the application of lasers in automated fiber/tape laying procedures. Laser systems with high power up to 10 kW are commercially available, and even more, if needed. Consider these variables when evaluating AFP diode laser sources:

Beam intensity (I) Power: The power of a diode laser module can be customized from just a few watts to 10s, even 100 s of kilowatts. For an AFP application, typically 1 kilowatt or less per tow is sufficient. The exact power depends mostly on the material and the processing speed.

Beam profile (Y) Beam homogeneity: The amount of heat, needs to be uniform across a designated target area.

Beam size and shape (W) Size and weight: A compact diode laser system allows for a more compact AFP head that can maneuver more intricate structures.

Beam angle (β) Incoming tape angle (α): When the laser beam is positioned toward the composite beam, part of the initial intensity (I0) will be reflected (Ir), partly absorbed (Ia), and partly transmitted (It). The amount of beam that is reflected, absorbed, or transmitted depends on a number of factors such as the characteristics of the material or strain, the wavelength of the beam, and the distribution of the fibers [4]. Ideally, the surface of the fiber-reinforced strip is covered with a thin and ideally flat layer of the thermoplastic matrix that aids in bonding. But in practice, such a layer rarely exists. The conical angle at which the radial rays fall on the material which is simultaneously under the pressure of the roller strikes the wedge-shaped cavity formed by the substrate and the thrust roller and causes different levels of reflection. The laser beam can be reflected from the thermoplastic matrix and also from the fibers on the surface of the tape. The part of the beam that is reflected from the surface should be as small as possible for laser heating to be effective. Any beam that is not reflected is transmitted or absorbed by the matrix and the fibers. It is actually the absorption of the beam that helps to heat the material.

2.5 Process parameters

Lee and Springer in their research [18] presented a model for the production of composites from thermoplastic composite tapes based on PEEK and carbon fiber using LATP/LAFP. They conclude that the process consists of three main steps—impregnation, consolidation, and crystallinity [18]. Impregnation is a parameter that cannot be controlled by automatic fiber/tape laying while consolidation and crystallinity can be controlled. Consolidation consists of two subprocesses—intimate contact and autohesion, that is direct connection or just connection (autohesion, healing). Autohesion is the formation of bonds between two surfaces of the identical polymer at elevated temperatures usually slightly above Tg. It is a new technique for precise bonding by self-bonding polymers without the need for adhesives. To achieve better final characteristics of the thermoplastic composite it is necessary to achieve a good degree of intimate contact and good self-connection.

Agrawal et al. [17] indicate that better on-site thermoplastic consolidation is achieved when a unified thermal model is applied by applying a wider beam with higher laser power. More authors [7, 38] also studied the effect of laser power, pressure, and deposition rate on the strength of the bond between layers. They found that the power of the laser had a dominant effect on the quality of the connection, and the force of the roller pressing had a minimal effect. Grove [36] in his research worked on the modeling of on-site consolidation processes and from experiments he concluded that the resulting poor crystallinity is due to the high cooling rate. Yousefpour and Ghasemi Nejhad [37] later proposed preheating the strip below the Tg glazing temperature to achieve better laminate consolidation.

2.6 Trial test with DOE

More and more factors have an influence on effectiveness and efficiency in industrial processes and systems. To find the optimum in control of the processes there are often a lot of experiments to realize—practical and theoretical ones. Design of experiments involve designing a set of experiments, in which all relevant factors are varied systematically (**Figure 6**). When the results of these experiments are analyzed, they help to identify optimal conditions.

In the previous chapter, we talked about some input parameters in LATP/LAFP process, and here in this chapter will talk about the three main output parameters that are determined after each experiment (trial test with DOE) in this process and that are—voids and crystallinity ILSS (**Figures 4** and **5**).

Voids: There are two types of voids that are inherent in the on-site consolidation process—intralaminar and interlaminar. The intralaminar cavity occurs during the impregnation of the tape, while, the interlaminar cavity is mainly the result of the process of laying the tape. The intralaminar cavity is embedded in the tape.

Crystallinity: Crystallinity also affects the mechanical properties of the final product and depends on the thermal cycle and cooling rate of the thermoplastic. In most LATP/LAFP applications, the cooling rate is as high as 1000°C per minute. Parts made of thermoplastic prepress tape based on PEEK and carbon fiber have a degree of crystallinity from 20 to 35% [18]. The two subprocesses—intimate contact and autohesia are a function of temperature, pressure, and consolidation time (passing speed), and these are the three most important parameters in any process. At a given pressure, a higher temperature and longer time are required to achieve the optimal degree of intimate contact and self-connection. Crystallinity affects the mechanical properties of thermoplastic composites. Higher crystallinity increases strength and rigidity [20]. On the other hand, the lower the degree of crystallinity increases the impact resistance and breaking strength [21–25]. The degree of crystallinity depends on the history of thermal processes. Low cooling rates result in a higher degree of crystallinity and vice versa. Hence, the degree of crystallinity achieved through on-site consolidation of the laying strip is limited due to the extremely high cooling rates. Kumar et al. [26] measured the degree of crystallinity of PEEK as well as the growth rate of spherolites (supermolecular forms of semicrystalline polymers) for samples heated to a melting point of 380°C and 420°C, then cooled to a crystallization temperature of 300°C or 320°C at a speed of about 3°C/s [26]. They concluded that the degree of crystallinity and the size of the crystals were higher for samples cooled by higher melting temperatures. In other words, at a constant cooling rate, the size of the polymer spherolites depends on the maximum processing temperature. Sonmez and Hahn [27] found that the degree of crystallinity of thermoplastic laminate obtained from one-way carbon fiber-based strips and REEK is between 25 and 35% in a single-site consolidation process. Similar degrees of crystallinity have been obtained in other studies indicating that PEEK is not sensitive to the cooling rates involved in the LATP/LAFP process [28, 29].

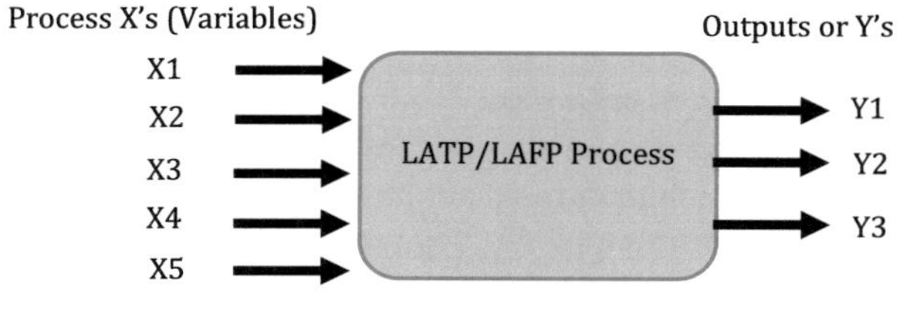

Figure 6.
Model of design of experiments.

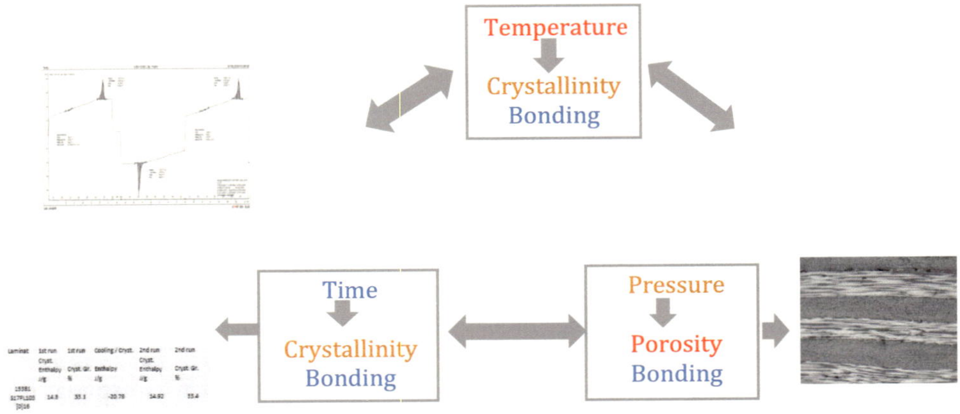

Figure 7.
Preliminary quality verification for final flat parts: Low porosity + Hight crystallinity + good bonding = good mechanical properties (tensile, flexural strength, ILSS, NOL test).

Mechanical properties: The most common mechanical tests performed to determine the process parameters are 3pbt and ILSS. These tests make it easier to see if the interlayers are well bonded during this process. If necessary, additional mechanical tests are performed. The ILSS survey is a well-established and fast calculation to be observed as a process (input parameters affect the final product).

2.7 Conclusion and recommendation for process parameters for the final product

Achieving complete consolidation by applying robotic processes requires the realization of additional research related to the optimization of many parameters for obtaining composite parts of different sizes and shapes. The interrelationship between the process parameters, the properties of the material, and the strength of the bond between the interlaminar layers has been investigated and shown in **Figure 7**.

Process parameters:

- Temperature (laser temperature, laser angle, mandrel's temperature)
- Pressure (compaction force of the roller)
- Time (layup speed) automatic tape laying process using a laser heating

3. Conclusion

- Robotic tape laying processes are very suitable for the efficient production of carbon fiber reinforced parts, especially for use in aerospace and other industries.
- The application of laser heating has some advantages over alternative heat sources, such as hot gas. In the hot gas torch (HGT) process, the thermal flux is difficult to control, while in the laser heating process, the laser beams are controlled spatially and temporally. Therefore, the obtained parts of composite materials with automated processes using laser application are expected to be of uniform quality, compared to those obtained by the HGT process. The two most important advantages of laser heating application are the high energy

density (accumulated energy per unit system per unit volume [J/m^3]) and the short response time. The first allows the application of higher speeds when laying lanes, while the second allows laying lanes of complicated geometries, including large variations in laying speed. However, the application of lasers also has its drawbacks. The cost of the equipment is high compared to conventional heat sources, and also the laser always requires the equipment to be housed in a protected environment. In addition, the application of laser heating requires a thorough understanding of the interaction of light with fiber-reinforced thermoplastic tapes.

- When the speed of laying the thermoplastic tapes is high, then the heat that enters from the laser source to the entrance to the material is negligible and insignificant. Therefore, the researchers concluded that a good model of heating of the thermoplastic material should be made in correlation with the angle at which the radial rays fall on it and the speed of placement of the laser in order to successfully connect the layers.

- Experiments [20–23] have shown that excellent interlaminar quality, and thus good properties of the final parts, can be obtained when the laying of the tape is at low speeds, higher compaction force, and laser beam in the case when it is primarily aimed at the tape.

Acknowledgements

Authors would like to acknowledge the support of the Engineering research team from Mikrosam D.O.O. and R&D team from the Institute for advanced composites and robotics from R.N. Macedonia.

Author details

Svetlana Risteska
Institute for Advanced Composites Robotics: IACR, Prilep, Republic of North Macedonia

*Address all correspondence to: svetlanar@iacr.edu.mk

References

[1] Orth T, Weimer C, Krahl M, Modler N. A review of radiative heating in automated layup and its modelling. Journal of Plastics Technology. 2017;**13**:92-125

[2] Beyeler E, Philips W, Guceri SI. Experimental investigation of laser-assisted thermoplastic tape consolidation. Journal of Thermoplastic Composite Materials. 1988;**1**:107-121. DOI: 10.1177/089270578800100109

[3] Shirinzadeh B, Foong CW, Tan BH. Robotic fibre placement process planning and control. Assembly Automation. 2000;**20**:313-320

[4] Soutis C. Fibre reinforced composites in aircraft construction. Progress in Aerospace Science. 2005;**41**:143-151

[5] Frketic J, Dickens T, Ramakrishnan S. Automated manufacturing and processing of fiber-reinforced polymer (FRP) composites. An additive review of contemporary and modern techniques for advanced materials manufacturing. Additive Manufacturing. 2017;**14**:69-86

[6] Crosky A, Grant C, Kelly D, Legrand X, Pearce G. Fibre placement processes for composite manufacture. advances in composites manufacturing and process design. Elsevier ltd.; 2015. pp. 79-92. https://doi.org/10.1016/B978-1-78242-307-2.00004-X

[7] Risteska S, Trajkovska Petkoska A, Samak S, Drienovsky M. Annealing Effects on the Crystallinity of Carbon Fiber-Reinforced Polyetheretherketone and Polyohenylene Laminate Composites Manufactured by Laser Automatic Tape Placement. 2020;**26**(3), Received 27 September 2018; Accepted 01 March 2019, ISSN1392-1320. Materials Science POLYMERS AND COMPOSITES. DOI: 10.5755/j01.ms.26.3.21489

[8] Khan MA, Mitschang P, Schledjewski R. Identification of some optimal parameters to achieve higher laminate quality through tape placement process. Advances in Polymer Technology. 2017;**29**(2):98-111. DOI: 10.1002/adv.20177

[9] Kassapoglou C. Design and Analysis of Composite Structures: With Applications to Aerospace Structures. 2nd ed. West Sussex: John Wiley & Sons Ltd.; 2013

[10] Mehran Eimanlou. Investigation of the effect of process parameters on bond strength of thermoplastic composite rings manufactured using fiber laser. A Master thesis in the Department of Mechanical, Industrial, and Aerospace Engineering: Québec, Canada: Concordia University Montréal; 2018.

[11] Grouve WJ, Warnet LL, Rietman B, Visser HA, Akkerman R. Optimization of the tape placement process parameter for carbon-PPS composites. Composites Part A: Applied Science and Manufacturing. 2013;**50**(1):44-53

[12] Khan MA, Mitschang P, Schledjewski R. Parametric study on processing parameters and resulting part quality through thermoplastic tape placement process. Journal of Composite Materials. 2012;**47**(4):485-499

[13] Stokes-Griffin CM, Compston P. The effect of processing temperature and placement rate on the short beam strength of carbon-fibre-PEEK manufactured using a laser tape placement process. Composites Part A Applied Science and Manufacturing. 2015;**78**:247-283

[14] Stokes-Griffin CM, Compston P. Laser-assisted tape placement of termoplastic composites: The effect of process parameters on bond strength. Sustainable Automotive Technologies 2013. Switzerland: Springer International

Publishing; 2014. DOI: 10.1007/978-3-319-01884-3_13

[15] Francesco R, Michael HS, Selpk IG. Effects of processing on laser assisted thermoplastic tape consolidation. Composites Part A Applied Science and Manufacturing. 1997;**28**(9):1023-1033

[16] Comer AN, Hammond PE, Ray DI, Lyons JO, Obande WI, Jones DA, et al. Wedge peel interlaminar toughness of carbon-fibre/PEEK thermoplastic laminates manufactured by laser-assisted automated-tape-placement (LATP). In: CETEC14; Finland. 2014. https://www.researchgate.net/publication/266558923_

[17] Agrawal V, Mccullough RL, Schultz JM. The thermoplastic laser-assited consolidation process - mechanical and microstructure characterization. Journal of Thermoplastic Composite Materials. 1996;**9**(4):365-380

[18] Lee WI, Springer GS. A model of the manufacturing process of thermoplastic matrix composites. Journal of Composite Materials. 1987;21:1017-1055

[19] Mark B, Gruber J, Tierney J, Jensen BJ et al. Thermoplastic in situ placement requires better impregnated tapes and tows. Baltimore, MD: Proceedings of the 2012 SAMPE conference and exhibition

[20] Gao SL, Kim JK. Cooling rate influences in carbon fibre/PEEK composites. Part I: Crystallinity and interface adhesion. Composites Part A: Applied Science and Manufacturing. 2000;31(6):517-530

[21] Samakoski B, Samak S, Risteska S, Sokoloski Z, Cvetkoska D. In situ consolidation for thermoplastic AFP process – Study on the input parameters for improvement of final product properties. 2019. The 14 th SAMPE CHINA Conference & Exhibition; May 6-8, 2019; Beijing; P.R. China: Oral SAMPE CHINA Conference & Exhibition 2019

[22] Samak S, Risteska S, Dukovski V, Trajkoski S. Some experimental investigation of products from thermoplastic composite materials manufactured with robot and LAFP. International Journal of Engineering Research & Technology. 2020;**9**(09):1082-1088

[23] Srebrenkoska S, Dukovski V, Risteska S. Influence of the process parameters on laser - assisted automated tape placement process. International Journal of Engineering Research & Technology. 2020;**9**(11):638-644

[24] Gao SL, Kim JK. Cooling rate influences in carbon fibre/PEEK composites. Part III: Impact damage performance. Composites Part A: Applied Science and Manufacturing. 2001;**32**(6):775-785

[25] Gao SL, Kim JK. Cooling rate influences in carbon fibre/PEEK composites. Part II: Interlaminar fracture toughness. Composites Part A: Applied Science and Manufacturing. 2001;**32**(6):763774

[26] Kumar S, Anderson DP, Adams WW. Crystallization and morphology of poly (aryl- ether-etherketone). Polymer. 1986;**27**(3):329-336

[27] Fazil O, Sonmez H, Hahn T. Analysis of the on-line consolidation process in thermoplastic composite tape placement. Journal of Thermoplastic Composite Materials. 1997;**10**:543-572

[28] Tierney J, Gillespie JW Jr. Modeling of heat transfer and void dynamics for the thermoplastic composite tow-placement process. Journal of Composite Materials. 2003;**37**(19):1745-1768

[29] Ranganathan S, Advani SG, Lamontia MA. A non-isothermal process model for consolidation and void

reduction during in-situ tow placement of thermoplastic composites. Journal of Composite Materials. 1995;**20**(8): 1040-1062

[30] Mantel S, Springer G. Manufacturing process models for thermoplastic composites. Journal of Composite Materials. 1992;**26**(16):2348-2377

[31] Khan MA, Mitschang P, Schledjewski R. Identification of some optimal parameters to achieve higher laminate quality through tape placement process. Advances in Polymer Technology. 2010;**29**(2):98-111

[32] Butler CA, Mccullough RL, Pitchumani R, Gillespie JW Jr. An analysis of mechanisms governing fusion bonding of thermoplastic composites. Journal of Thermoplastic Composite Materials. 1998;**11**(4):338-363

[33] Ageorges C, Ye L, Mai YW, Hou M. Characteristics of resistance welding of lap shear coupons: Part II. Consolidation. Composites Part A: Applied Science and Manufacturing. 1998;**29**(8):911-919

[34] Stokes Griffin CM, Compston P. Investigation of sub-melt temperature bonding of carbonfiber/PEEK in an automated laser tape placement process. Composites Part A: Applied Science and Manufacturing. 2016;**84**:17-25

[35] Yassin K, Hojjati M. Processing of thermoplastic matrix composites through automated fiber placement and tape laying methods. A review. Journal of Thermoplastic Composite Materials. 2018;**31**(12)

[36] Grove S. Thermal modelling of tape laying with continuous carbon fibre reinforced thermoplastic. Composites. 1988;**19**(5):367-375

[37] Yousefpour A, Ghasemi Nejhad MN. Experimental and computational study of APC- 2/AS4 thermoplastic composite C-rings. Thermoplastic Composite Material. 2001;**14**(2):129-145

[38] Samak S, Sokoloski Z, Risteska S, Bogdanoski D. Improving the final properties of thermoplastic composites manufactured with laser automated tape placement (LATP). In: The 13th SAMPE CHINA Conference & Exhibition 2018; Shanghai, P. R. China. 2018

Chapter 4

Mechanically Improved and Multifunctional CFRP Enabled by Resins with High Concentrations Epoxy-Functionalized Fluorographene Fillers

Junhua Wei

Abstract

To meet the maximum potential of the mechanical properties of carbon fiber reinforced plastics (CFRP), stress transfer between the carbon fibers through the polymer matrix must be improved. A recent promising approach reportedly used reinforcing particles as fillers dispersed in the resin. Carbon based fillers are an excellent candidate for such reinforcing particles due to their intrinsically high mechanical properties, structure and chemical nature similar to carbon fiber and high aspect ratio. They have shown great potential in increasing the strength, elastic modulus and other mechanical properties of interest of CFRPs. However, a percolation threshold of ~1% of the carbon-based particle concentration in the base resin has generally been reported, beyond which the mechanical properties deteriorate due to particle agglomeration. As a result, the potential for further increase of the mechanical properties of CFRPs with carbon-based fillers is limited. We report a significant increase in the strength and elastic modulus of CFRPs, achieved with a novel reinforced thermoset resin that contains high loadings of epoxy-reacted fluorographene (ERFG) fillers. We found that the improvement in mechanical performance of CFRPs was correlated with increase in ERFG loading in the resin. Using a novel thermoset resin containing 10 wt% ERFG filler, CFRPs fabricated by wet layup technique with twill weaves showed a 19.6% and 17.7% increase in the elastic modulus and tensile strength respectively. In addition, because of graphene's high thermal conductivity and high aspect ratio, the novel resin enhanced CFRPs possessed 59.3% higher through-plane thermal conductivity and an 81-fold reduction in the hydrogen permeability. The results of this study demonstrate that high loadings of functionalized particles dispersed in the resin is a viable path towards fabrication of improved, high-performance CFRP parts and systems.

Keywords: CFRP, functionalized graphene, tensile properties, hydrogen permeability, through-plane thermal conductivity

1. Introduction

Carbon Fiber Reinforced Plastics (CFRP) have been widely used as lightweight materials replacing metals because of their superior specific strength, stiffness, and corrosion resistance [1]. In recent years, their application has been extended from high-end industries, like aerospace, aircraft, sports, and military to cost-sensitive industries, like automotive and energy [2]. This new trend requires reducing the cost of parts made from CFRP. This price reduction can be achieved through two approaches: 1) enhancing mechanical properties of CFRPs resulting in lower usage for same function and 2) eliminating functional parts by enabling CFRPs with these functions.

Improving the stress transfer between the carbon fibers (CF) and improvement of stiffness of polymer matrix are most popular research topics. Besides development of highly efficient CFRP fabrication techniques, low price CF manufacturing and material-saving structure designs, these studies also aim at cost reduction via decrease in CFRP usage. Due to the weak interfacial interaction between the load bearing CFs and the polymer matrix that integrate the structure, delamination is one of the major failure modes for CFRP. One of the key reasons for this failure is that CF has relatively low density of surface functional groups, which result in lesser surface interactions. To address this problem, increasing the interface interaction through sizing, a polymer coating that improves stress transfer through chemical bond and physical interlock, is required as part of the CF manufacturing process [3]. Inspired by the sizing mechanism, rigid particles have been added either into polymer sizing or by spraying on the CFs, to encourage strong physical interlock.

New approach to lower the cost and weight of systems containing CFRPs is to make CFRP a functional part in the system. For example, Type V storage vessels used for hydrogen storage are designed with only a single layer of CFRP which endures the compressed gas pressure and simultaneously acts as an efficient hydrogen barrier. Compared with other tank designs (e.g. Type IV), Type V tank is more cost competitive via process, material, and weight saving through limiting the hydrogen barrier liner. However, to realize this design, a large amount of fillers with good gas barrier properties need to be present into the CFRP to form an effective gas barrier layer.

Carbon-based materials use in polymer sizing have drawn a lot of attention to this approach because of their superior mechanical properties, chemical structure like that of CFs, and most importantly due to their high aspect ratio. Methods including electrophoretic deposition [4, 5], spray [6, 7], and dip coating [8–12] have been employed to deposit graphene and carbon nanotubes onto CFs surfaces. CFRPs made with these coated CFs show ~20–70% improvement on interlaminar shear strength. Up to 10 wt% of particles laden sizing coated onto CF by such approaches have been demonstrated [8]. However, this approach generally does not improve the tensile or flexural modulus [5, 8, 13]. This is because the polymer matrix is much weaker compared to CF, and as a result, the polymer matrix deforms under stress. To reinforce both the modulus and strength of CFRP, researchers have tried to add carbon-based particulate materials directly into the resin matrix as well, [14–19] but this approach achieved only limited improvement in mechanical properties. The primary limitation is particle aggregation, which prevents high particle loading resins [16, 20]. Typically, CFRP with ~1 wt% of carbon-based fillers in the polymer matrix have shown the best properties. A polymer matrix containing a high loading of dispersed carbon-based materials would therefore be likely to radically increase both, strength and modulus, of CFRPs. Meanwhile, carbon-based materials have been appraised for their excellent properties as electrical and thermal conductors, gas barrier, fire retardant et al. [21, 22]. The advantages of their multifunction can introduce CFRP with special properties to replace other functional part and reduce the device price, when the particle concentrations are high enough.

Fluorographene (FG) is a unique carbon-based material which maintains the excellent mechanical properties of graphene but with excellent chemical, electronic,

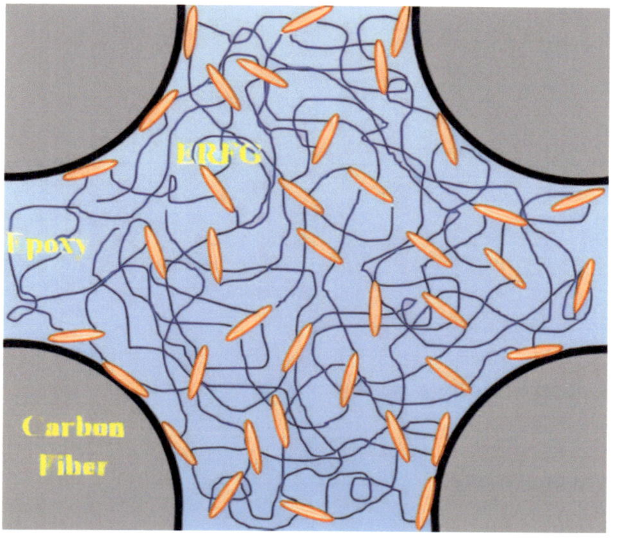

Figure 1.
Illustration of CFRP containing a high loading of epoxy-reacted fluorographene (ERFG, orange ovals) in the polymer matrix.

tribological, and hydrophobic properties, due to the presence of fluorine atoms [23]. One distinctive advantage of FG is that it is easily forms dispersion and exfoliates because fluorination disrupts the van der Waals forces between FG sheets [24]. Other researchers have studied the electrical, thermal and mechanical properties of FG-reinforced polyimide [25–27]. We have previously reported on dispersing epoxy-monomer–functionalized FG in epoxy with 30 wt% loading, which achieved ~90%, ~60%, and ~ 170% improvement in the tensile strength, modulus, and toughness respectively [28]. In this work, resin containing ERFG was used to make CFRP by wet lay-up method. The high particle loading in the polymer matrix improves stress transfer between the CF, both by physically interlocking with the CFs to improve strength, and by chemical linkage within the polymer matrix to improve modulus (**Figure 1**). Additionally, the high particle loading also enables the CFRP to exhibit properties of FG present within the CFRP, such as low gas permeability, high thermal conductivity, et al. This mechanically improved and multifunctional CFRP is a very competitive composite for cost sensitive CFRP applications.

2. Experimentation

2.1 Materials

Graphite fluoride (FG, also called fluorographite), triethylenetetramine (TETA), and acetone were bought from Sigma Aldrich. Bisphenol A based epoxy resins (EPON 826) with 178 g/mol EEQ (Epoxy Equivalent Weight [29]) was purchased from Miller-Stephenson Chemical Co., Inc. Fibre Glast 2000 (Epoxy), Fibre Glast 2060 (Hardener), 2 K 2 × 2 twill weave carbon fiber fabrics and all the materials for wet layup were from Fibre Glast. Toray T700 12 K 2 × 2 plain weave fabrics and T300 3 K 2 × 2 plain weave fabrics were purchased from Composite Envisions. T800S 12 K 2 × 2 twill weave fabrics and M55JB 6 K 2 × 2 plain weave fabrics were supplied by Rock West Composites. MGS L285 (epoxy) and H287 (hardener) were purchased from Aircraft Spruce & Specialty Co.

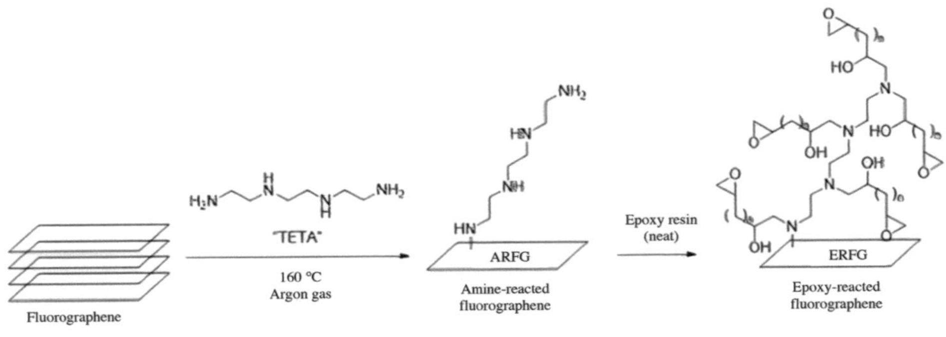

Figure 2.
Synthesis of ARFG and ERFG from fluorographite.

2.2 ERFG particle synthesis

ERFG particles were synthesized using the same experimental procedure as in our earlier published work [28, 30], and is presented in **Figure 2**. Briefly, fluorographite was dispersed and solvent exfoliated in TETA, then the C-F bonds on the fluorographite were substituted with amine groups from TETA at high temperature under argon gas protection, thereby producing amine-reacted fluorographene (ARFG). The ARFG particles were recovered from the TETA via filtration, washed, and then dispersed into epoxy resin (EPON 826), to synthesize epoxy-monomer-reacted fluorographene (ERFG).

2.3 CFRP fabrication

The formulation was prepared by a two-step procedure. Firstly, ERFG particles were added into an epoxy resin by centrifugal mixing. Immediately before use for CFRP fabrication, the epoxy resin containing ERFG particles was mixed with a hardener. The optimum weight ratio between the epoxy/ERFG mixture to the hardener was determined by a dynamic DSC screening for the maximum curing enthalpy. The CFRP panels were fabricated according to literature [31]. Next, the formulation consisting of epoxy with ERFG particles and hardener was spread onto each layer of CF fabric. After piling and spreading, accessories like the vacuum bagging were added. The impregnated CF fabrics were cured at room temperature for 5 hr. under vacuum, and then post-cured at 100°C for another 5 hr. The cured CFRP was removed from vacuum and bagging and was then cut into specific shapes according to ASTM D3039 by a water-cooled diamond saw for tensile testing.

2.4 Characterization

The chemical functionalization of the particles was characterized by an ATR FT-IR (Perkin Elmer). The morphology of the ERFG particles was imaged by a TEM (Jeol 1200).

The mechanical properties of CFRP were measured according to ASTM 3039 using a universal tester (AGS-X, Shimadzu). The through-plane thermal conductivity was measured as per ASTM D7984 using a thermal conductivity analyzer (TCi, C-Therm). Hydrogen permeability tests were performed by a professional testing lab (Versaperm Ltd.) in accordance with ASTM G148.

3. Results

3.1 Particle characterization

The mechanism of attaching molecular amine groups to fluorographene through substitution of surface fluorine atoms has been previously reported [32]. In this work, TETA served both as solvent to exfoliate fluorographite and as a reagent for introduction of amino functional groups onto the exfoliated fluorographene to produce Amine Reacted Florographene (ARFG) particles. The color of fluorographite changed from white to black during this substitution reaction. In the second step, the amine groups on ARFG were capped with EPON 826, a low molecular weight diglycidyl epoxy resin derived from bisphenol A. FT-IR spectroscopy was used to identify the chemical groups present on the particles during each step of synthesis. In **Figure 3a**, a prominent feature at 1260 cm^{-1} characteristic of covalent C-F stretching of fluorographite was detected [33]. There are three identifying absorptions peaks for TETA: 1050–1200 cm^{-1}, 1450 cm^{-1}, 1600 cm^{-1} representing C-N stretching, C-H bending, and N-H bending, respectively [34, 35]. In the ARFG FTIR spectrum, a broad absorption peak appears from 1000 cm^{-1} to 1200 cm^{-1} as a result of C-N stretching appeared in place of C-F stretching. The C-F stretching absorption peak at 1260 cm^{-1} is greatly diminished in comparison to the starting material, fluorographite. The 1580 cm^{-1} absorption on the ARFG is inherited from TETA representing N-H bending. In the FTIR spectrum ofERFG, the epoxy functionality display a large number of additional absorptions peaks: 1034 cm^{-1}, 1184 cm^{-1} and 1241 cm^{-1}, 1300 cm^{-1}, 1509 cm^{-1} representing C-O-C ether stretching, C-C-O-C stretching, stretching of cyclic ether, and C-C stretching of aromatic ring, respectively [36, 37]. In ERFG, the C-O-C absorptions extended the broad feature seen in ARFG (1000 cm^{-1} to 1200 cm^{-1}) to a broader range (1000 cm^{-1} to 1300 cm^{-1}), which indicates the presence of epoxy groups onto the ERFG particles. The high aspect ratio, single-digit microsized, and few-layer sheet [38] ERFG was shown in the TEM micrographs of ERFG particles (**Figure 3b**).

3.2 CFRP characterization

CFRPs were fabricated with Fibre Glast CF fabrics and Fibre Glast 2000/2060 resin base with different weight loadings of ERFG according to the procedure

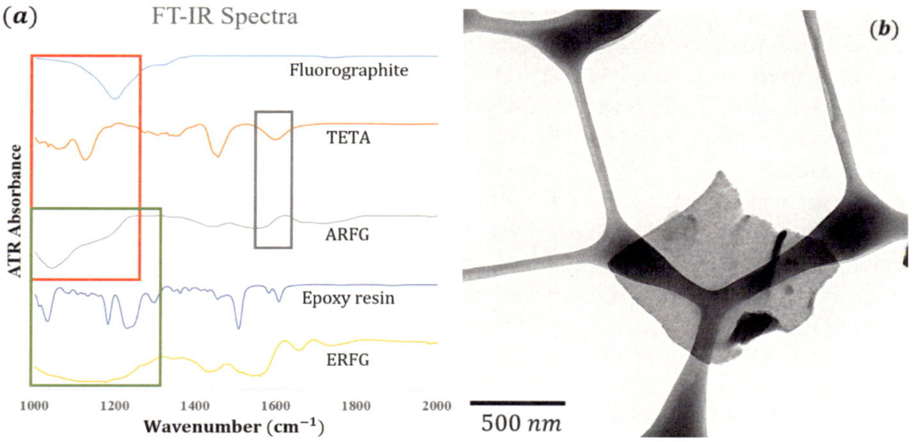

Figure 3.
FT-IR spectra of ERFG and its synthetic precursors (a), and TEM image of an ERFG particle (b).

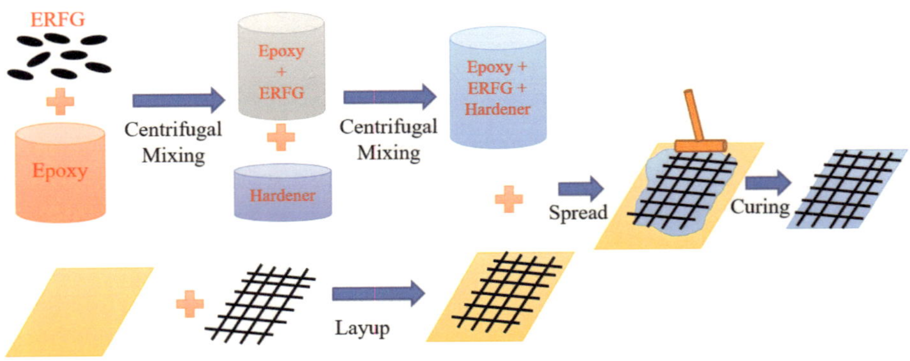

Figure 4.
Fabrication of CFRP using wet layup method.

Sample	Pure (0 wt% ERFG)	5 wt% ERFG	10 wt% ERFG	15 wt% ERFG
Elastic Modulus (GPa)	51.5 (1.3)	55.1 (7.6)	61.6 (4.4)	63.9 (3.1)
Tensile Strength (MPa)	500.7 (15.6)	536.3 (14.9)	608.9 (22.8)	583.9 (22.0)
Thermal conductivity (W/(m × k))	0.425 (0.005)	0.578 (0.001)	0.677 (0.006)	NA
Hydrogen permeability (mol/(m × s × Pa))	1.79E-14 (0.01)	NA	2.20E-16 (0.02)	NA

Note: the carbon fibers take 37 vol% to 42 vol% in the tested CFRPs.

Table 1.
Mechanical properties, through-plane thermal conductivity, and hydrogen permeability of CFRP made by Fibre Glast CF fabrics and Fibre Glast 2000/2060 resin with different wt% of ERFG (standard deviation).

shown in **Figure 4**. The mechanical, thermal, and gas barrier properties of these samples were investigated and are summarized in **Table 1**.

Unlike the carbon fiber surface particle coating from previous reports, adding particles to the polymer matrix increased both the elastic modulus and the tensile strength. The increase of the elastic modulus of the CFRP made by ERFG resin correlated well with the increase of the ERFG loading below 15 wt%. However, the tensile strength decreased for CFRP with 15 wt% ERFG. There are two possible explanation for this phenomenon. First, the viscosity of resin with ERFG particles increased when adding more particles. For highly viscous resin, higher vacuum pressure and elevated temperature are required for uniform wetting. Second, 15 wt% of particles in the resin can form aggregates or liquid crystal domains. Both these structures prevent the uniform distribution of the particles through the CF fabric. Since the enhanced interfacial interaction between CFs and polymer matrix is based on the premise of attachment of the ERFG present in the polymer matrix to the surface of CF to form a physical interlock. The formation of aggregates or liquid crystal domains decreases the density of ERFG coated onto the CF surface and causes less effective stress transfer and results in decrease of CFRP's overall tensile strength. Further investigations are underway to reveal which mechanism dominates at higher particle loading CFRPs.

The in-plane thermal conductivity of C FRP can reach ~1000 W/(m × K) due to the continuous nature of the carbon fibers [39]. In contrast, the through-plane thermal conductivity of CFRP is in the 0.2–0.5 W/(m × K) range [40] because the polymer matrix separates the carbon fiber tows. Historically reported approaches

Material	Hydrogen permeability 10E-16 mol/(m × s × Pa)	Ref.
ERFG-enhanced CFRPs	2.20	This work
HDPE	11.20–8.09	[44, 45]
LDPE	24.50	[45]
PA	6.69–2.23	[45]
PTFE	36.20	[46]

Table 2.
Hydrogen permeability of conventional polymer liner for COPV tanks [43].

added highly thermally conductive particles like graphene [40] or diamond [41] into the polymer matrix to increase the through-plane conductivity, but this compromises the other properties of CFRP, mainly mechanical properties. In this work, a significant improvement in the through-plane conductivity is achieved. The ERFG particles serve as bridges to facilitate the conduction of heat between individual carbon fibers. A 60% improvement in the through-plane thermal conductivity was realized in this way (**Table 1**). This significant improvement on the through-plane thermal conductivity can improve the CFRP's stability under thermal cycling and will enable the replacement of metal built electronic box in weight essential applications [42].

The 2D structure and high aspect ratio of the ERFG particles dispersed within the cured epoxy matrix were efficient in blocking the through-plane permeation of hydrogen gas. The hydrogen permeability of CFRP made with 10 wt% ERFG was 81 times lower than that of CFRP made with pure, unfilled epoxy resin (**Table 2**). These results make ERFG-enhanced CFRPs as state-of-the-art hydrogen barrier materials, far exceeding the properties of conventional polymer liner used in COPV (Composite Overwrapped Pressure Vessels) tanks [43, 47, 48]. The decreased hydrogen permeation is a huge benefit for building COPV tanks for hydrogen or natural gas storage. Additionally, use of ERFG-enhanced CFRPs eliminates one manufacturing step – polymer liner pre-molding – required to manufacture the current Type IV COPV tanks. This multifunctional CFRP is very promising for direct manufacturing of liner-less Type V COPV tanks.

4. Discussion

In this work, the idea of adding high loadings of graphene derivative into the polymer matrix to increase both the elastic modulus and tensile strength of CFRP has been achieved with Fibre Glast fabrics and 2000/2060 resin. To further demonstrate the universality of this idea, we tested the tensile properties of CFRP made with another widely used resin system, MGS 285/287, and with different classes of CFs. To maintain consistency, all CF fabrics used were made with Toray CF. In these carbon fibers, T300 and T700 represent the standard modulus CF. T300 is primarily used for automotive parts and sports products, and T700 is used in gas tanks. The T800 represents the intermediate modulus carbon fiber used in aircraft applications. The M55 is a high modulus carbon fiber mainly used in spacecraft. MGS resin systems have been widely used in CFRP wet layup fabrication with applications in aircraft. By testing CFRP made using widely popular commercial resin with different grades of CFs, we demonstrated the universality of our idea.

Table 3 summarizes the elastic modulus and tensile strength of CFRPs made with different CF using MGS 285/287 resin with 10 wt% ERFG. the elastic modulus

CF	Elastic Modulus (GPa)		Tensile Strength (MPa)	
	Pure (0 wt% ERFG)	10 wt% ERFG	Pure (0 wt% ERFG)	10 wt% ERFG
T300	45.9 (3.4)	51.5 (0.9)	402.9 (14.2)	432.0 (25.1)
T700	51.9 (3.7)	59.1 (7.1)	586.9 (26.7)	701.8 (48.0)
T800	56.8 (3.1)	65.2 (4.7)	683.7 (45.9)	762.5 (37.3)
M55	94.5 (2.0)	111.8 (2.5)	376.2 (28.7)	427.1 (38.0)

Note: the carbon fibers take 38 vol% to 45 vol% in tested CFRPs.

Table 3.
Mechanical properties of CFRP made by different types of CF fabrics and 10 wt% ERFG MGS resin.

of all the CFRP samples made with ERFG show a 15–20% increase in these mechanical properties, due to formation of highly uniform dispersion of ERFG in the polymer matrix and from stronger interfacial interaction between polymer and ERFG through epoxy monomer functionalization. However, the trend observed in tensile strength is more complex. With T700, T800, and M55 CFRPs, the tensile strength increased by ~15–20% when 10 wt% ERFG was added in the resin. When compared, the strength of CFRPs with T300 CF increased by only ~10%. The variation with use of different carbon fibers could be the result of different sizing used on the carbon fibers and structure of fabrics. The thickness, stiffness, and chemical structure of the sizing radically affects the interface between carbon fiber and the polymer matrix. The addition of ERFG amplifies the effect of sizing on interfacial interactions. Different CF weaving structure and fabric density (gsm) affect the resin penetration through the fabric, which together with the increased resin viscosity due to dispersed ERFG particles, resulted in a non-uniform distribution of the particles across CFRP thickness.

5. Conclusions

In this work, significant improvements to both the elastic modulus and the tensile strength of CFRP were achieved through high ERFG particle loadings in the polymer matrix of CFRP. The low surface tension of C-F groups in the fluorographite precursor material, the epoxy monomer functionalization, and the small size, collectively result in high loading of well-dispersed exfoliated ERFG in epoxy resin. Furthermore, the loading of ERFG is high enough to enhance other properties of CFRP composite. The through-plane thermal conductivity of CFRP was increased by ~60% and the hydrogen permeability of CFRP was decreased by a factor of ~81. This work provides a universal method to reduce the cost of CFRPs by 1) decreasing the weight of CFRP parts through improved mechanical properties, and by 2) reducing system complexity by having CFRP parts multiple desirable properties which would ordinarily require separate parts made from different materials. The approach reported in our work represents a viable strategy for low cost parts made with CFRP for applications in automotive, sporting goods and energy sectors.

Author details

Junhua Wei
PARC, A Xerox Company, Palo Alto, CA, USA

*Address all correspondence to: jawei@parc.com

References

[1] C. Soutis, *Fibre reinforced composites in aircraft construction*. Progress in Aerospace Sciences, vol. 41, no. 2, pp. 143-151, 2005/02/01/ 2005.

[2] *Carbon Fiber Market by Raw Material (PAN, Pitch, Rayon), Fiber Type (Virgin, Recycled), Product Type, Modulus, Application (Composite, Non-composite), End-use Industry (A & D, Automotive, Wind Energy), and Region - Global Forecast to 2029*. MARKETSANDMARKETS2019.

[3] Y. Luo, Y. Zhao, Y. Duan, and S. Du, *Surface and wettability property analysis of CCF300 carbon fibers with different sizing or without sizing.* Materials & Design, vol. 32, no. 2, pp. 941-946, 2011/02/01/ 2011.

[4] T. Sun, M. Li, S. Zhou, M. Liang, Y. Chen, and H. Zou, *Multi-scale structure construction of carbon fiber surface by electrophoretic deposition and electropolymerization to enhance the interfacial strength of epoxy resin composite.s* Applied Surface Science, vol. 499, p. 143929, 2020/01/01/ 2020.

[5] C. Wang, J. Li, S. Sun, X. Li, F. Zhao, B. Jiang, and Y. Huang, *Electrophoretic deposition of graphene oxide on continuous carbon fibers for reinforcement of both tensile and interfacial strength*. Composites Science and Technology, vol. 135, pp. 46-53, 2016/10/27/ 2016.

[6] A. K. Srivastava, V. Gupta, C. S. Yerramalli, and A. Singh, *Flexural strength enhancement in carbon-fiber epoxy composites through graphene nano-platelets coating on fibers.* Composites Part B: Engineering, vol. 179, p. 107539, 2019/12/15/ 2019.

[7] F. Awan and T. Subhani, *Preparation and Characterization of Carbon Nanotube Deposited Carbon Fiber Reinforced Epoxy Matrix Multiscale Composites.* Advanced Nano Research, vol. 1, no. 1, p. 14, 2017.

[8] X. Zhang, X. Fan, C. Yan, H. Li, Y. Zhu, X. Li, L. and Yu, *Interfacial Microstructure and Properties of Carbon Fiber Composites Modified with Graphene Oxide.* ACS Applied Materials & Interfaces, vol. 4, no. 3, pp. 1543-1552, 2012/03/28 2012.

[9] W. Qin, C. Chen, J. Zhou, and J. Meng, *Synergistic Effects of Graphene/ Carbon Nanotubes Hybrid Coating on the Interfacial and Mechanical Properties of Fiber Composites.* Materials (Basel, Switzerland), vol. 13, no. 6, p. 1457, 2020.

[10] L. Liu, F. Yan, M. Li, M. Zhang, L. Xiao, L. Shang, and Y. Ao, *Improving interfacial properties of hierarchical reinforcement carbon fibers modified by graphene oxide with different bonding types.* Composites Part A: Applied Science and Manufacturing, vol. 107, pp. 616-625, 2018/04/01/ 2018.

[11] J. Fu, M. Zhang, L. Jin, L. Liu, N. Li, L. Shang, M. Li, L. Xiao and Y. Ao, *Enhancing interfacial properties of carbon fibers reinforced epoxy composites via Layer-by-Layer self assembly GO/SiO2 multilayers films on carbon fibers surface.* Applied Surface Science, vol. 470, pp. 543-554, 2019/03/15/ 2019.

[12] C. Xiao, Y. Tan, X. Wang, L. Gao, L. Wang, and Z. Qi, *Study on interfacial and mechanical improvement of carbon fiber/epoxy composites by depositing multi-walled carbon nanotubes on fibers.* Chemical Physics Letters, vol. 703, pp. 8-16, 2018/07/01/ 2018.

[13] A. Ashori, H. Rahmani, and R. Bahrami, *Preparation and characterization of functionalized graphene oxide/carbon fiber/epoxy nanocomposites.* Polymer Testing, vol. 48, pp. 82-88, 2015/12/01/ 2015.

[14] Y. Li, H. Zhang, Z. Huang, E. Bilotti, and T. Peijs, *Graphite Nanoplatelet*

Modified Epoxy Resin for Carbon Fibre Reinforced Plastics with Enhanced Properties. Journal of Nanomaterials, vol. 2017, p. 5194872, 2017.

[15] S. S. Reddy, C. Yuvraj, and K. P. Rao, *Design, Analysis, Fabrication and Testing of CFRP with CNF Composite Cylinder for Space Applications.* International Journal of Composite Materials, vol. 5, no. 5, p. 2166, 2015.

[16] E. Bekyarova, E. Thosternson, A. Yu, M. Itkis, D. Fakhrutdinov, T. Chou, and R. Haddon, *Functionalized Single-Walled Carbon Nanotubes for Carbon Fiber–Epoxy Composites.* The Journal of Physical Chemistry C, vol. 111, no. 48, pp. 17865-17871, 2007/12/01 2007.

[17] S. Aldajah and Y. Haik, *Transverse strength enhancement of carbon fiber reinforced polymer composites by means of magnetically aligned carbon nanotubes.* Materials & Design, vol. 34, pp. 379-383, 2012/02/01/ 2012.

[18] T. Glaskova-Kuzmina, A. Aniskevich, M. Zarrelli, A. Martone, and M. Giordano, *Effect of filler on the creep characteristics of epoxy and epoxy-based CFRPs containing multi-walled carbon nanotubes.* Composites Science and Technology, vol. 100, pp. 198-203, 2014/08/21/ 2014.

[19] A. K. Pathak, M. Borah, A. Gupta, T. Yokozeki, and S. R. Dhakate, *Improved mechanical properties of carbon fiber/graphene oxide-epoxy hybrid composites.* Composites Science and Technology, vol. 135, pp. 28-38, 2016/10/27/ 2016.

[20] J. Kim, J. Cha, B. Chung, S. Ryu, and S. H. Hong. Composites Science and Technology, vol. 192, p. 108101, 2020/05/26/ 2020.

[21] J. Wei and J. Qiu, *Allyl-Functionalization enhanced thermally stable graphene/fluoroelastomer nanocomposites.* Polymer, vol. 55, no. 16, pp. 3818-3824, 2014/08/05/ 2014.

[22] Y. Cui, S. I. Kundalwal, and S. Kumar, "Gas barrier performance of graphene/polymer nanocomposites," *Carbon,* vol. 98, pp. 313-333, 2016/03/01/ 2016.

[23] J. Robinson, J. Burgess, C. Junkermeier, S. Badescu, T. Reinecke, F. Perkins, M. Zalalutdniov, J. Baldwin, J. Culbertson, P. Sheehan, and E. Snow, *Properties of Fluorinated Graphene Films.* Nano Letters, vol. 10, no. 8, p. 3001, 2010.

[24] X. Ye, P. Gong, J. Wang, H. Wanga, S. Ren, and S. Yang, *Fluorinated graphene reinforced polyimide films with the improved thermal and mechanical properties.* Composites: Part A, vol. 75, p. 96, 2015.

[25] Z. Chen, H. Huang, S. Yan, Z. Zheng, S. Liu, Y. Yuan, J. Zhao, and Y. Fu, *New Synthetic Approach of Fluorine-Containing Graphene Oxide for Improving Dielectric and Mechanical Properties of Polyimide Composites.* Ind. Eng. Chem. Res., vol. 26, p. 9926, 2017.

[26] W. Li, W. Zhao, L. Mao, S. Zhou, C. Liu, Z. Fang, and X. Gao, *Investigating the fluorination degree of FG nanosheets on the tribological properties of FG/PI composite coatings.* Progress in Organic Coatings, vol. 139, p. 105481, 2020.

[27] X. Yin, Y. Feng, Q. Zhao, Y. Liu, S. Li, H. Dong, W. Hu, and W. Feng, *Highly transparent, strong, and flexible fluorographene/fluorinated polyimide nanocomposite films with low dielectric constant.* Journal of Materials Chemistry C, vol. 6, p. 6378, 2018.

[28] E. Beh, G. Iftime, J. L. B. Rivest, J. Wei, and R. Pandey, *Composite Materials Comprising Chemically Linked Fluorographite-Derived Nanoparticles*. US, 2019, Application number: 16/225125.

[29] E. P. Douglas, H. R. Hamilton, J. C. Nino, A. Stewart, and J. Tatar, *Highly*

accelerated lifetime for externally applied bond critical fiber-reinforced polymer (FRP) infrastructure materials. Department of Materials, Science & Engineering University of Florida, 2014.

[30] J. Wei, G. Iftime, J. L. B. Rivest, and S. Garner, *Fiber Reinforced Plastic Enhanced by Functionalized Graphene.* US, 2018, Application number 16/228456.

[31] J. S. Tomblin, Y. C. Ng, and K. S. Raju, *Material Qualification and Equivalency for Polymer Matrix Composite Material Systems: Updated Procedure.* DOT/FAA/AR-03/19, 2003.

[32] H. Barès, A. Bakandritsos, M. Medved, J. Ugolotti, P. Jakubec, O. Tomanec, S. Kalytchuk, R. Zboril, and M. Otyepka, *Bimodal role of fluorine atoms in fluorographene chemistry opens a simple way toward double functionalization of graphene*. Carbon, vol. 145, pp. 251-258, 2019/04/01/ 2019.

[33] K.-J. Jeon, Z. Lee, E. Pallak, L. Moreschini, A. Bostwick, C. Park, R. Mendelsberg, V. Radmilovic, R. Kostecki, T. Richardson, and E. Rotenberg, *Fluorographene: A Wide Bandgap Semiconductor with Ultraviolet Luminescence.* ACS Nano, vol. 5, no. 2, pp. 1042-1046, 2011/02/22 2011.

[34] Z. ALOthmana, E, Yilmazb, M. Habilla, I. Alsohaimia, A. Aldawsaria, N. Al-Harbic, and M. Soylak, *Triethylenetetramine modified multiwalled carbon nanotubes for the efficient preconcentration of Pb(II), Cu(II), Ni(II) and Cd(II) before FAAS detection.* RSC Advances, vol. 5, pp. 106905-106911, 2015.

[35] N. Ngamkitpinyo and A. Imyim, *Selective extraction of Pb(II) using triethylenetetramine-modified polystyrene-divinylbenzene resin.* J. Mater. Environ. Sci., vol. 6, no. 5, pp. 1562-1569, 2015.

[36] S. Jahanshahi, A. Pizzi, A. Abdulkhani, and A. Shakeri, *Analysis and Testing of Bisphenol A—Free Bio-Based Tannin Epoxy-Acrylic Adhesives.* Polymers, vol. 8, p. 143, 2016.

[37] R. A. Chowdhury, M. Hosur, M. Huruddin, A. Tcherbi-Narteh, A. Kumar, V. Boddu, and S. Jeelani, *Self-healing epoxy composites: preparation, characterization and healing performance.* Journal of Materials Research and Technology, vol. 4, no. 1, pp. 33-43, 2015/01/01/ 2015.

[38] N. Liaros, A. B. Bourlinos, R. Zboril, and S. Couris, *Fluoro-graphene: nonlinear optical properties.* Opt. Express, vol. 21, pp. 21027-21038, 2013.

[39] C. A. Silva, E. Marotta, M. Schuller, L. Peel, and M. O'Neill, *In-Plane Thermal Conductivity in Thin Carbon Fiber Composites.* Journal of Thermophysics and Heat Transfer, vol. 21, no. 3, pp. 460-467, 2007.

[40] F. Wang and X. Cai, *Improvement of mechanical properties and thermal conductivity of carbon fiber laminated composites through depositing graphene nanoplatelets on fibers.* Journal of Materials Science, vol. 54, no. 5, pp. 3847-3862, 2019/03/01 2019.

[41] M. Srinivasan, P. Maettig, K. Glitza, B. Sanny, A. Schumacher, M. Duhovic, and J. Schuster, *Out of Plane Thermal Conductivity of Carbon Fiber Reinforced Composite Filled with Diamond Powder.* Open Journal of Composite Materials, vol. 6, pp. 41-57, 2016.

[42] M. Martins, R. Gomes, L. Pina, C. Pereira, O. Reichmann, D. Teti, N. Correia, and N. Rocha, *Highly Conductive Carbon Fiber-Reinforced Polymer Composite Electronic Box: Out-Of-Autoclave Manufacturing for Space Applications.* Fibers and Polymers, vol. 6, p. 92, 2018.

[43] N. L. Newhouse, *Development of Improved Composite Pressure Vessels for Hydrogen Storage. Hexagon Lincoln*

Department of Energy Efficiency and Renewable Energy Fuel Cell Technologies Office 2016.

[44] P. D. L. Staff, W. Woishnis, Ed. *Permeability and Other Film Properties of Plastics and Elastomers*. Plastics Design Library, 1995.

[45] H. Saechtling and F. Pabst, *Kunststoff-Taschenbuch* (no. v. 24). Hanser, 1989.

[46] D. O. Stodilka, N. P. Kherani, W. T. Shmayda, and S. J. Thorpe, *A tritium tracer technique for the measurement of hydrogen permeation in polymeric materials.* International Journal of Hydrogen Energy, vol. 25, no. 11, pp. 1129-1136, 2000/11/01/ 2000.

[47] J. Humpenöder, *Gas permeation of fibre reinforced plastics.* Cryogenics, vol. 38, no. 1, pp. 143-147, 1998/01/01/ 1998.

[48] D. Schultheiß, *Permeation Barrier for Lightweight Liquid Hydrogen Tanks*. PhD, Mathematisch-Naturwissenschaftlichen, Universit¨at Augsburg, Publikationsserver der Universit¨at Augsburg, 2007.

Section 3

Natural Fiber Reinforced

Chapter 5

Composite Materials with Natural Fibers

Nicholas Lambrache, Ora Renagi, Lidia Olaru and Brian N'Drelan

Abstract

The materials involved in the fabrication of biocomposites have dissimilar physical and chemical properties. More important, the newly created materials exhibit anisotropy and their performance is strongly influenced by the hydrophobic nature of the natural fibers used as reinforcement materials. Beyond a compressive discussion regarding the potential of composite materials with natural fibers in engineering applications, the chapter focuses on simulation of their behavior under applied loads. Modern experimental approaches for defining and validating computer simulations are also introduced. Finally, health hazards and biodegradability issues are evaluated. The new trends in biocomposites materials for engineering applications are briefly discussed.

Keywords: biocomposites classification, anisotropy, hydrophoby, simulation and experimental trends, health hazards

1. Introduction

Environmental and economic considerations recommend the employment of natural fibers as reinforcements in polymeric matrices. The matrix protects the fibers from environmental degradation and by doing so it preserves its mechanical strength. Natural fibers have lower density than artificial fibers, are a renewable material and in most cases are recyclable. The absolute tensile strength of natural fibers is lower than the tensile strength of artificial fibers, as shown in the schematic representation from **Figure 1**. However, due to their lower density, natural fibers have a higher specific tensile strength – defined as the ratio between the absolute tensile strength and the mass density – which makes them ideal candidates for composites employed in aerospace applications.

There are important variations in the physical and chemical properties of natural fibers, variations depending on age, geographical location and age. Such variations must be considered in any design and evaluation of composites with natural fibers and statistical evaluations of strength are important.

Natural fibers are hydrophobic and anisotropic. Such properties pose challenges to the designer and manufacturer of biocomposites. The need to control the content of water in natural fibers requires specific chemical treatment in order to maximize the bond fiber-matrix. The mechanical properties of biocomposites need to be rigorously evaluated by precision experiments and statistical evaluations designed

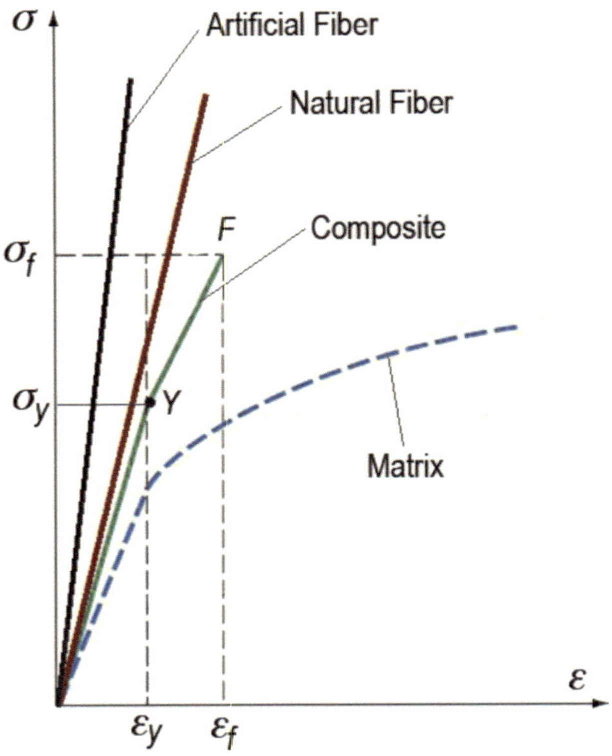

Figure 1.
Stress–strain plots for composites.

to assure reliable input parameters for the computer modeling and simulation of components made from such materials.

This chapter introducer the reader to key aspects encountered in the classification of natural fibers. The performance evaluation, benefits and disadvantages of designing, manufacturing and employment of biocomposites in a wide range of industrial applications are also discussed in some detail.

2. Classification of natural fibers

Natural fibers are the result of geological processes or are produced in the bodies of plants and animals. A general but not exhaustive evaluation of such fibers as potential reinforcing materials in composites is based on their origin:

- Fibers resulted from geological processes. A notable example is asbestos.
- Fibers of animal origin. Such fibers include silk, wool, mohair and alpaca fiber.
- Plant fibers. Such fibers include seed fibers, leaf fibers, bast fibers, fruit fibers and stalk fibers.

2.1 Fibers resulted from geological processes

Asbestos is the only natural fiber produced by geological processes widely used as reinforcement in composite materials. It is a fibrous silicate mineral. There are six types of such fibrous minerals, including amosite, crocidolite, tremolite, chrysotile, vermiculite and actinolite. All types of fibrous silicate minerals belonging to asbestos family are composed of long, thin crystals, each one composed of a multitude of microscopic fibrils that are easily released into atmosphere. Their inhalation triggers lung health conditions including mesothelioma, asbestosis and cancer, see [1]. For such reasons composites with asbestos are completely banned in many countries. One characteristic sample of chrysotile asbestos from the biggest asbestos mine in the world – Jeffrey Mine, Quebec, Canada, is shown in **Figure 2**. Chrysotile asbestos belongs to the family of serpentine minerals.

2.2 Animal fibers

The most commonly used animal fibers are the wool of domestic sheep and the silk. Other common animal fibers are alpaca fibers and mohair from Angora goats. The silk fiber is secreted by glands located near the mouth of some specific insects during the preparation of their cocoons.

All animal fibers contain specific proteins such as collagen, keratin and fibroin. Collagen is the main structural protein in connective tissues and the most abundant protein in mammals, see [2]. Keratin belongs to the family of structural proteins and is a key structural component of hair, horns, claws or skin in vertebrates, see [3]. In silk, keratin confers excellent tensile strength to the fiber, in the range 650 MPa–750 MPa, see [4]. Regardless of their great importance in bioscience, both wool and silk are less used in the fabrication of composite materials for industrial applications compared with plant fibers.

2.3 Plant fibers

Plant fibers can be collected from leaves, bast, fruit, seeds or stalks. Common leaf fibers include abaca and sisal. Leaf fibers are characterized by increased strength, most likely due to their high content of lignin, see [4]. The most remarkable type of fruit fiber is the coir of coconuts. Bast fibers are part of the outer cell layers of stems of plants and known examples include Kenaf and hemp. Specific plant fibers used as reinforcement in composite materials:

- Bast Fibers: Flax, Kenaf, Jute, Hemp, Pandanus
- Leaf Fibers: Abaca, Pineapple, Banana, Sisal

Figure 2.
Chrysotile from Jeffrey mine, Quebec, Canada. Author's collection.

- Fruit Fibers: Coir
- Stalk Fibers: Straws of Wheat, Rice, Bamboo

Plant fibers offer significant advantages over synthetic fibers as reinforcing components. They are renewable, biodegradable, have low densities and lower processing costs. However, they also bring design and fabrication challenges, mostly due to their reduced adhesion to the polymer matrices as a consequence of their hydrophilic character, see [5]. Some images of plants and the related fibers are shown below. The roots of Pandanus – *Pandanus Utilis* – form a pyramidal tract and it holds the trunk of the plant. Pandanus trees grow in tropical and sub-tropical coastlines and islands of all oceans and can withstand salt spray, drought and strong winds, see [6, 7]. The plants growing along seashores have thick aerial roots as anchors in the sand. Such roots keep the trees upright and secure them in the ground - see **Figure 3**.

The Coconut Palmtree is known for its abundance in the equatorial and tropical regions of the world and also for its versatility – see **Figure 4**. The *Cocos Nucifera* or Coconut Palmtree belongs to the family *Arecaceae*, genus *Cocos* [8]. The drupes are used for food, charcoal, oil and cosmetics, and also in the fabrication of biofuel and composite materials. The mechanical strength of the husk fibers is investigated in many research centers around the world [9–12] and the results show widely distributed values.

The coir or coconut fiber is extracted from the outer husk of the fruit. It has a low density, is unsinkable and resistant to saltwater. The individual fiber cells are narrow and hollow, with thick walls made of cellulose – see **Figure 5**. They are pale when immature, but later become hardened and yellowed as a layer of lignin is

Figure 3.
Pandanus tree from rainforest habitat, PNG University of Technology campus, Lae, Morobe, Papua New Guinea. Inlet: Pandanus Fibers.

Figure 4.
Coconut Palmtree from PNG University of Technology campus, Lae, Morobe, Papua New Guinea. Inlet: Coir.

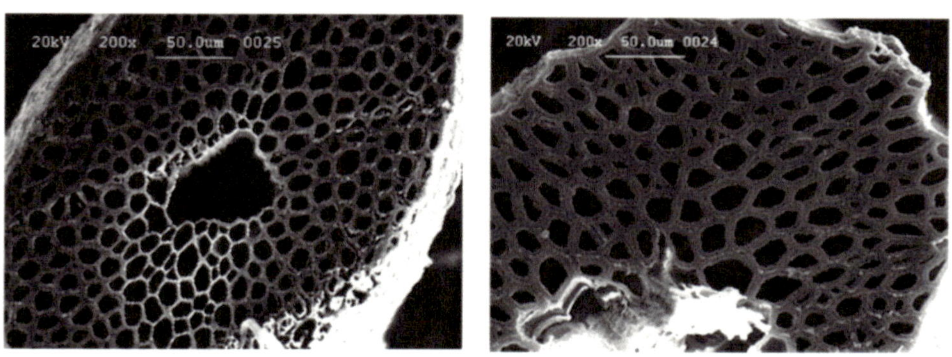

Figure 5.
Cross sections of coir - left - and sisal Fibers – Right – at 200X magnification. Scan electron microscopy on SEMoscope Inovenso IEM 11.

deposited on their walls. Each cell is about 1 mm (0.04 in) long and 10 to 20 μm in diameter. Fibers are typically 100 to 200 mm long. The two varieties of coir are brown and white. Brown coir harvested from fully ripened coconuts is thick, strong and has high abrasion resistance.

Sisal - botanical name *Agave Sisalana* - is a flowering plant native to Yucatan but widely naturalized in many tropical and subtropical regions of the world. The fibers are extracted from the sword-shaped leaves – see **Figure 6**. During its life of 8 to 10 years, the plant grows hundreds of leaves, each one containing around one thousand fibers [13]. Sisal has been the leading material for binder twine for centuries due to its strength, durability, ability to stretch, and resistance to deterioration

Figure 6.
Agave Sisalana from PNG University of Technology campus, Lae, Morobe, Papua New Guinea. Inlet: Sisal Fibers.

in saltwater. However, its traditional use is limited by the competition from polypropylene and the increased potential as reinforcement in composite materials.

The most important constituents of plant fibers are cellulose, hemicellulose, pectin and lignin. Cellulose is an organic polymer with chemical formula $(C_6H_{10}O_5)_n$. The high tensile strength of plant stems arises from the arrangement of cellulose fibers and their distribution into the lignin matrix. Cellulose is responsible for the hydrophilic nature of plant fibers. Hemicellulose is a heteropolymer present along cellulose in all terrestrial plant cell walls, see [14]. Hemicellulose has a random, amorphous structure with little strength, is not resistant to hydrolysis and can be easily hydrolyzed by dilute acid or base. Pectin has the function of holding plant fibers together. Lignin is an organic polymer made by cross-linking phenolic precursors, see [15]. It allows the development of structural materials in the cell walls, lending them strength. In adition, lignin is resistant to decay and its contribution to water absorption is negligible. At low magnifications on scan electron microscopes cross sections of coir and sisal fibers reveal similar honeycomb structures constituted of cellulose fibers in lignin matrices, see **Figure 5**.

3. Technology trends on composite materials with natural fibers

The resins used in manufacturing composite materials can be thermosetting and thermoplastic. The thermosetting resins are predominant, while the thermoplastic resins play a minor role in the fabrication of advanced composites.

The thermosetting resins require the addition of a curing hardener. Once cured, the composite cannot be reformed. Widely used thermosetting resins include epoxies, polyurethanes, phenolic and amino resins and polyamides. The most common employed are the epoxies because the potential for respiratory exposure is low due to their relatively high molecular-weight.

Another essential component of the cured resins is the hardener. Such compounds control the reaction rate and greatly influence the properties of the matrix. Some commonly used curing agents are the aromatic amines. Like epoxies, such hardeners have a low vapor pressure and are not a respiratory hazard. However, both epoxies and hardeners are a dermal hazard and can even permeate common protective gloves.

The authors of this chapter performed extensive experiments with epoxy resins and hardeners from Struers GmbH, Hanover, Germany. Their epoxy resins are suitable for the matrix of composite materials because they have low shrinkage and the adhesion to natural fibers is excellent. The hardened epoxy is duro-plastic and not affected by moderate heat or chemicals. The curing time is relatively long, but adhesion to most materials is excellent. They polymerize through a chemical reaction after being mixed in the correct proportions. Our experiments employed the polymer Epiclorhydrin in a ratio 15/2 with the hardener Triethylenetretamine. The hardened epoxy is duro-plastic and not affected by moderate heat or chemicals [16].

The curing of epoxy matrix depends on the amount of resin. If small amounts of resin are employed, the polymerization takes longer. However, favorable conditions exist for removing excessive heat generated by the chemical reaction. Larger amounts of epoxy will accelerate the curing process by storing heat due to the poor conductive properties of the system. The air bubbles developed during the curing are caused by higher than acceptable temperatures. Sticky or rubbery sample surfaces after curing indicate a process temperature too low and can be corrected by post-curing in ovens at temperatures below 50°C. The content of water in natural fibers can significantly influence the bond fiber-matrix and it can be reduced by combined exposure to heat at around 120°C for 2 hours and immersion in 10% NaOH for 4 hours. The immersion of fibers in NaOH is increasing the exposed surface, with beneficial effects for their adhesion to the matrix material.

4. Stress behavior simulation of composite materials with natural fibers

All major software platforms for Computer Aided Design and Finite Element Method Simulations allow the study of composites with natural fibers via the Composite Shell implementation. There are three types of composite options to define the arrangement of plies, thicknesses, material properties, and orientations, see **Figure** 7.

- An asymmetric laminate has an asymmetric arrangement of plies about the mid-surface. This is the most general composite option. A schematic of an asymmetric laminate with five layers is shown. Different colors represent different material properties and orientations. The shell mesh is created at the mid-plane.

- A symmetric laminate has a symmetric arrangement of plies - materials, ply orientations, and thicknesses - about the mid-surface. This implies symmetric ply thicknesses, material properties, and material orientations about the mid-plane.

- The sandwich composite is a particular case symmetric laminates with three layers. Such laminates are employed when higher resistance to bending loads is required. The outer two plies are recommended to be stiffer, stronger, and thinner than the middle ply. The core is usually lighter to reduce the overall

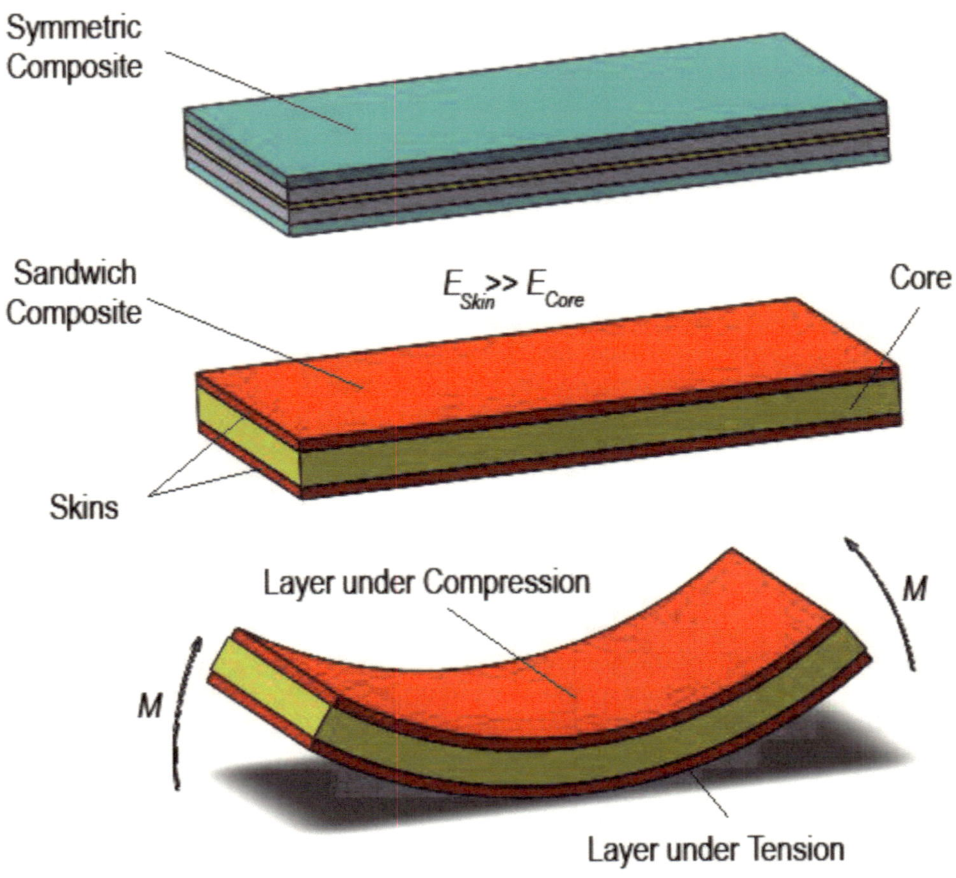

Figure 7.
3D CAD models of symmetric and Sandwich composites.

mass of the composite and has a high shear modulus of elasticity to resist the sliding of the skins.

Composite shells are used for structural members with weight savings required. The shells offer improved fatigue strength, corrosion resistance, and thermal conductivity. In the Shell Property Manager of such applications it is possible to define symmetric, asymmetric or sandwich composites with a number of layers up to 50. The Property Manager also allows users to set different thicknesses, material orientations and same or different material properties for each ply.

Composite laminates are special orthotropic materials that are modeled as single surfaces with several layers of orthotropic materials. A cross-ply sisal fiber-reinforced polymer is an example of a composite laminate with orthotropic material properties for each ply. A rock is an example of an orthotropic material that does not qualify as a composite. Notable differences between a composite laminate and an orthotropic material body:

Composite Laminate:

- Uses a laminated shell element formulation. In addition to other stress results, displays interlaminar shear stress between two adjacent plies. Delamination can occur between two plies with high stress values.

- Uses these unique failure criteria: Tsai-Hill, Tsai-Wu and Maximum Stress

Orthotropic Material Body:

- Uses an element formulation appropriate to the selected body. Interlaminar shear stress components do not apply for bodies defined as orthotropic materials.

- Uses the following failure criteria: Maximum von Mises Stress, Maximum Shear Stress / Tresca, Mohr-Coulomb Stress and Maximum Normal Stress

4.1 Fundamental stress analysis concepts

All system designers must, at early stages, identify the probable failure modes, select a suitable parameter by which severity of loading and environment may be analytically represented, propose a material and geometry for components and implement critical strength properties related to the probable failure mode. The magnitude of the loading severity parameter must be calculated under applicable loading and environmental conditions, and compared with the critical strength property. Failure may be averted by assuring that the loading severity parameter is safely less than the corresponding critical strength property for each potential failure mode.

The most important loading severity parameters are stress, strain, and strain energy per unit volume. Of these, stress is usually selected for calculation purposes. To completely define the state of stress at any selected point within a solid body, it is necessary to describe the magnitudes and directions of stress vectors on all possible planes that could be passed through such point. One way of defining the state of stress at a point is to determine all stress components that can occur on the faces of an infinitesimal cube of material placed at the origin of an arbitrarily selected right-handed Cartesian coordinate system of known orientation. Each of these components of stress may be classified as either a normal stress σ normal to a face of the cube, or a shear stress τ parallel to a face of the cube. The illustration in **Figure 8** depicts all possible stress components acting on an infinitesimal cubic volume element of dimensions *dx-dy-dz*.

Depending upon whether its material behaves in a *brittle* or a *ductile* manner, failure at the governing critical point of a component is dependent upon the principal normal stresses, the principal shearing stresses, or some combination of these. In any event, the designer must evaluate principal normal stresses and principal shearing stresses for any combination of applied loads. To do this, the general stress cubic Eq. (1) may be employed to find the principal stresses $\sigma_1, \sigma_2, \sigma_3$ and as a function of the readily calculable components of stress $\sigma_x, \sigma_y, \sigma_z, \tau_{xy}, \tau_{yz}, \tau_{zx}$ relative to any selected *x-y-z* coordinate system. The general stress cubic equation, developed from equilibrium concepts, has the form – see [17]:

$$\begin{aligned}\sigma^3 - \sigma^2(\sigma_x + \sigma_y + \sigma_z) + \sigma\left(\sigma_x\sigma_y + \sigma_y\sigma_z + \sigma_z\sigma_x - \tau_{xy}^2 - \tau_{yz}^2 - \tau_{zx}^2\right) - \\ -\left(\sigma_x\sigma_y\sigma_z + 2\tau_{xy}\tau_{yz}\tau_{zx} - \sigma_x\tau_{yz}^2 - \sigma_y\tau_{zx}^2 - \sigma_z\tau_{xy}^2\right) = 0\end{aligned} \tag{1}$$

Since all normal and shearing stress components are real numbers, all three roots of the general stress cubic equation are real. These three roots are the principal normal stresses $\sigma_1, \sigma_2, \sigma_3$. It is also possible to find the directions of principal stress vectors and principal shearing stress vectors if necessary. Furthermore, it can be shown that the magnitudes of the principal shearing stresses may be calculated from Eq. (2):

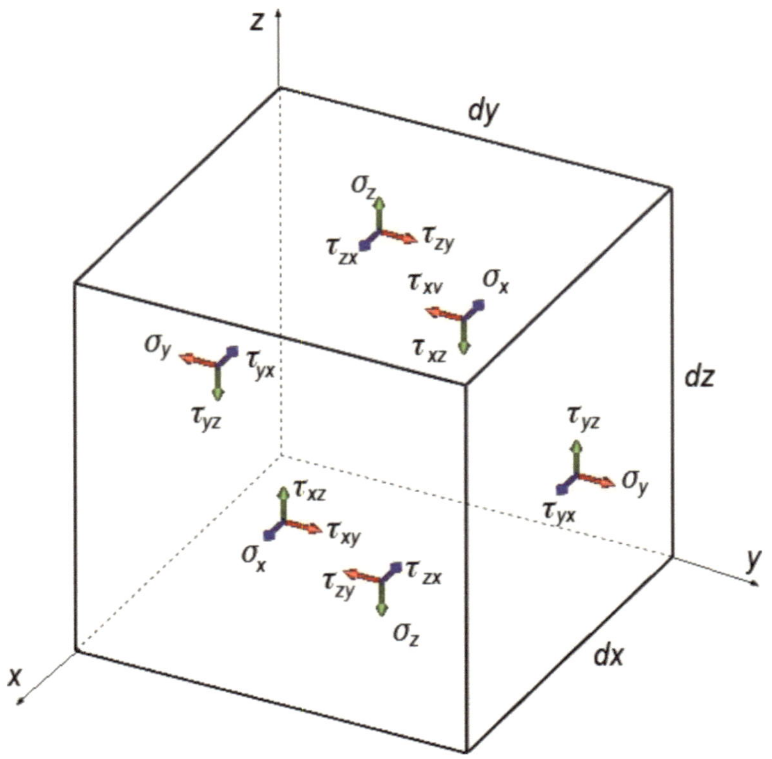

Figure 8.
Complete definition of the state of stress at arbitrary points.

$$|\tau_1| = \left|\frac{\sigma_2 - \sigma_3}{2}\right|, \quad |\tau_2| = \left|\frac{\sigma_3 - \sigma_1}{2}\right|, \quad |\tau_3| = \left|\frac{\sigma_1 - \sigma_2}{2}\right| \tag{2}$$

To summarize, if loads and geometry are known for a component in any system, the designer may identify the critical point, arbitrarily select a convenient x-y-z coordinate system, and calculate the resultant six stress components $\sigma_x, \sigma_y, \sigma_z, \tau_{xy}, \tau_{yz}, \tau_{zx}$. The above equations can be solved to find the principal normal stresses and the principal shearing stresses.

4.2 Failure criteria for composite materials

To determine whether a laminate will fail due to any applied load, the stresses across the different plies needs to be calculated and next a failure criterion based on these stress levels must be selected. The failure of composites occurs in multiple steps. When the stress in the first ply or first group of plies is high enough, it fails. This point of failure is the first ply failure, beyond which a laminate can still carry the load. For a safe design, laminates should be exposed to stresses below this point. The point where the total failure occurs is termed the ultimate laminate failure. Failure of composites occurs on a micromechanical scale due to fiber damage, matrix cracking, or interface or inter-phase failure. The local failure modes mentioned above cannot predict global laminate failure satisfactorily. Composite failure theories predict global laminate failure. These failure theories can be interactive, non-interactive or partially interactive. The non-interactive theories do not consider the interaction between different stress components, whereas the interactive theories do. The three theories available for laminate failure criteria are:

- Maximum Stress Criterion
- Tsai-Hill Failure Criterion
- Tsai-Wu Failure Criterion

The *Maximum Stress Criterion* widely applies to composite shells. Failure occurs according to the maximum stress criterion when the stress in one of the principal material directions exceeds the strength in that direction. The overall state of stress in the global coordinates is first computed by the program. Then, the program computes stress along the principal material directions for each lamina by applying a coordinate transformation. The program assumes a state of plane stress - 2D - for a lamina with $\sigma_3 = 0, \tau_{13} = 0, \tau_{23} = 0$. The failure index *FI* is computed as follows:

$$FI = \max\left(\frac{\sigma_1}{S_1}, \frac{\sigma_2}{S_2}, \left|\frac{\sigma_{12}}{S_{12}}\right|\right) \tag{3}$$

The software reports a Factor of Safety $FOS = \frac{1}{FI} = \frac{1}{\max\left(\frac{\sigma_1}{S_1}, \frac{\sigma_2}{S_2}, \left|\frac{\sigma_{12}}{S_{12}}\right|\right)}$. The composite will not fail if $FOS > 1$. The significance of the parameters involved in the equations above is:

- S_1 - Tensile strength of laminate in direction 1
- S_2 - Tensile strength of laminate in direction 2
- S_{12} - Shear strength of laminate

There are some additional requirements:

- If $\sigma_1 > 0$, $S_1 = S_{1\ Tensile}$
- If $\sigma_1 < 0$, $S_1 = S_{1\ Compressive}$
- If $\sigma_2 > 0$, $S_2 = S_{2\ Tensile}$
- If $\sigma_2 < 0$, $S_2 = S_{2\ Compressive}$

The maximum stress criterion has the following characteristics:

- It does not consider the interactions between different stress components into account as seen from the above equations.
- Predicts specific failure modes since stress in each material principal direction is compared with strength in that direction.

The *Tsai-Hill Criterion* considers the distortion energy portion of the total strain energy that is stored due to loading. The distortion energy is the portion of strain energy that causes shape change. The other portion is the dilatation energy that causes volume or area change due to loading. For composite shells each lamina is assumed to be in a state of plane stress with $\sigma_3 = 0, \tau_{13} = 0, \tau_{23} = 0$. The failure index is computed as:

$$FI = \frac{\sigma_1^2}{S_1^2} - \frac{\sigma_1 \sigma_2}{S_1^2} + \frac{\sigma_2^2}{S_2^2} + \frac{\tau_{12}^2}{S_{12}^2} \tag{4}$$

The significance of the parameters involved in Eq. (4) above is:

- S_1 - Tensile strength of laminate in direction 1
- S_2 - Tensile strength of laminate in direction 2
- S_{12} - Shear strength of laminate

The program reports a Factor of Safety $FOS = \frac{1}{\sqrt{FI}} = \frac{1}{\sqrt{\frac{\sigma_1^2}{S_1^2} - \frac{\sigma_1 \sigma_2}{S_1^2} + \frac{\sigma_2^2}{S_2^2} + \frac{\tau_{12}^2}{S_{12}^2}}}$.

The *FOS* is the coefficient by which all stress components should be multiplied to reach laminate failure at $FI = 1$. The composite will not fail if $FOS > 1$. There are also some additional requirements for this failure criterion:

- If $\sigma_1 > 0$, $S_1 = S_{1\ Tensile}$
- If $\sigma_1 < 0$, $S_1 = S_{1\ Compressive}$
- If $\sigma_2 > 0$, $S_2 = S_{2\ Tensile}$
- If $\sigma_2 < 0$, $S_2 = S_{2\ Compressive}$

The Tsai-Hill criterion considers the interaction between different stress components. However, it cannot predict various failure modes including fiber failure, matrix failure, and fiber-matrix interface failure.

The *Tsai-Wu Criterion* is applied to determine the factor of safety for composite orthotropic shells. This criterion considers the total strain energy - both distortion energy and dilatation energy - for predicting failure. It is more general than the Tsai-Hill failure criterion because it distinguishes between compressive and tensile failure strengths. For a 2D state plane stress - $\sigma_3 = 0$, $\tau_{13} = 0$, $\tau_{23} = 0$ - the Tsai-Wu failure criterion is expressed as:

$$F_1\sigma_1 + F_2\sigma_2 + 2F_{12}\sigma_1\sigma_2 + F_{11}\sigma_1^2 + F_{22}\sigma_2^2 + F_6\tau_{12} + F_{66}\tau_{12}^2 = 1 \tag{5}$$

The coefficients F_{ij} of the orthotropic Tsai-Wu failure criterion are related to the material strength parameters of the lamina determined by experiments. They are calculated from with following equations:

$$F_1 = \left(\frac{1}{S_{1\ Tensile}} - \frac{1}{S_{1\ \mathrm{Compressive}}}\right), F_2 = \left(\frac{1}{S_{1\ Tensile}^2} - \frac{1}{S_{1\ Compressive}^2}\right),$$

$$F_{12} = -\frac{1}{2}\sqrt{\frac{1}{S_{1\ Tensile}^2 \cdot S_{1\ Compressive}^2} \cdot \frac{1}{S_{2\ Tensile}^2 \cdot S_{2\ Compressive}^2}},$$

$$F_{11} = \frac{1}{S_{1\ Compressive}^2 \cdot S_{1\ Tensile}^2}, F_{22} = \frac{1}{S_{2\ Compressive}^2 \cdot S_{2\ Tensile}^2},$$

$$F_6 = \left(\frac{1}{S_{12\ T}^2} - \frac{1}{S_{12\ C}^2}\right), F_{66} = \frac{1}{S_{12\ C} \cdot S_{12\ T}} \tag{6}$$

The significance of the parameters involved in the equations above is:

- $S_{1\ Tensile}$ - Tensile strength of laminate along fiber direction
- $S_{1\ \text{Compressive}}$ - Compressive strength of laminate along fiber direction
- $S_{2\ Tensile}$ - Tensile strength of laminate transverse to fiber direction
- $S_{2\ \text{Compressive}}$ - Compressive strength of laminate transverse to fiber direction
- $S_{12\ T}$ - Positive shear strength of laminate
- $S_{12\ \text{C}}$ - Negative shear strength of laminate, considered equal to the positive shear strength by the solver in Finite Element Analysis. Directions 1 and 2 refer to the fiber direction, respectively transversal to it.

The factor of safety *FOS* is the coefficient by which all laminate stress components should be multiplied to reach laminate failure according to the Tsai-Wu criterion stated above. The *FOS* for laminate failure is calculated as:

$$FOS = \frac{R - C_1}{2C_2}, C_1 = F_1\sigma_1 + F_2\sigma_2 + F_6\tau_{12},$$

$$C_2 = F_{11}\sigma_1^2 + F_{22}\sigma_2^2 + F_{66}\tau_{12}^2 + 2F_{12}\sigma_1\sigma_2, R = \sqrt{\left|C_1^2 + 4C_2\right|} \tag{7}$$

If $FOS > 1$, the composite will not fail. The Tsai-Wu failure criterion cannot predict different failure modes including fiber failure, matrix failure, and fiber-matrix interface failure.

4.3 Example

Consider a composite material with natural fibers. The composite has seven plies, a layup [0/90/0/90/0/90/0] and it is subject to three point bending with support span of 20 mm by applying a uniform load P = 10 N/mm at the center. The objective of the simulation is to evaluate the longitudinal stress SX, the displacement UY at point B and the transverse shear stress TXY at point C. The geometric and mechanical properties of the composite as evaluated by measurement and experiments are:

- Thickness of Pliers 1, 2, 3, 5, 6, 7: d = 0.1 mm
- Thickness of Ply 4: d = 0.4 mm
- The fibers on plies are oriented normal to each other, see **Figure 9**.
- Model: Linear Elastic Orthotropic
- Elastic Modulus: $E_X = 1e + 11\ N/m^2$, $E_Y = 5e + 9\ N/m^2$, $E_Z = 5e + 9\ N/m^2$
- Poisson's Ratio: $\nu_{XY} = 0.4$, $\nu_{YZ} = 0.3$, $\nu_{XZ} = 0.3$
- Shear Modulus: $G_{XY} = 3e + 9\ N/m^2$

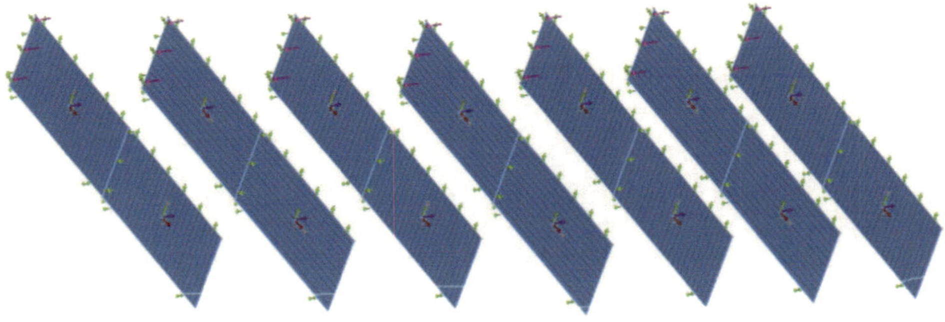

Figure 9.
Plies orientation.

- Mass Density: $\rho = 1400\ kg/m^3$
- Tensile Strength in X: $S_e = 1.4256e + 8\ N/m^2$
- Yield Strength: $S_y = 1.3904e + 8\ N/m^2$
- Default Failure Criterion: Max von Mises Stress [17, 18]
- Due to symmetry, only a quarter of the composite is modeled

The simulation results are shown in **Figure 10** and are validated by experiments.

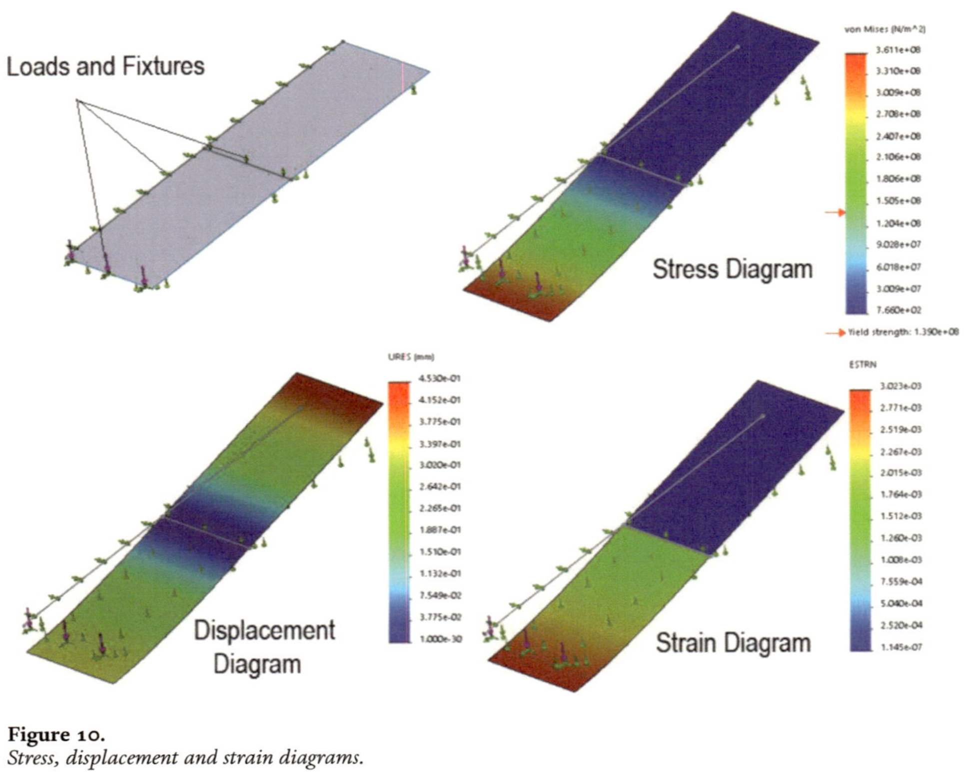

Figure 10.
Stress, displacement and strain diagrams.

5. Modern experimental approaches

Laboratory-scale tensile and torsion testing machines allow experimental evaluations between any applied tensile/torque loads and the induced linear elongations or twist on composite specimens with natural fibers. In engineering terms, they facilitate the experimental evaluation of yield strength, Young modulus E or shear modulus G with accuracy. The specimen can also be brought to the point of destruction and such experiments allow the establishment of the failure point of the material.

Such experiments can be performed with the required precision on SM1002 Bench-Top Tensile Testing Machines and SM1001 Torsion Testing Machines from Tecquipment. The experimental evaluation of Poisson's Ratio with the relationship $\nu = -d\varepsilon_t/d\varepsilon_a$ performed under EU Standard 10,002–1 directions is also important because it provides input parameters for simulations.

If the fibers are randomly oriented in the matrix may be acceptable to consider the composite a homogenous, isotropic linear elastic material. Under such

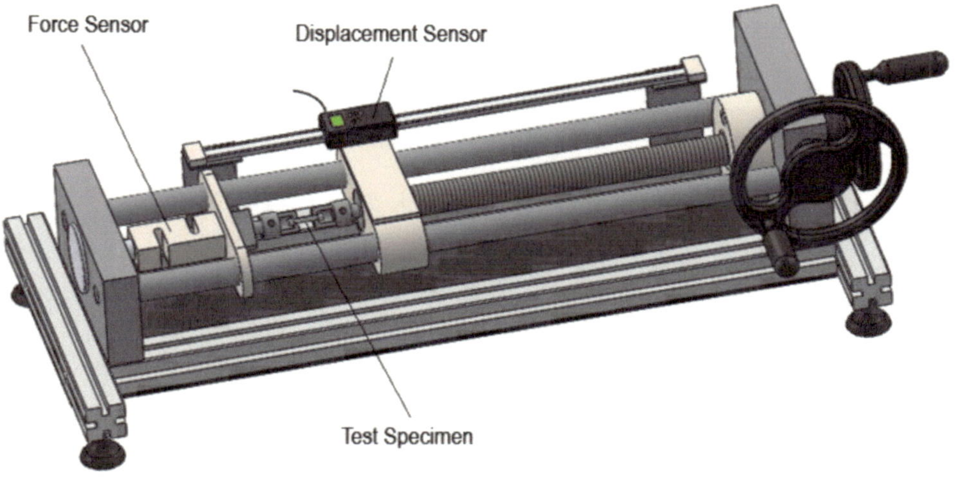

Figure 11.
3D CAD model of tensile testing machine.

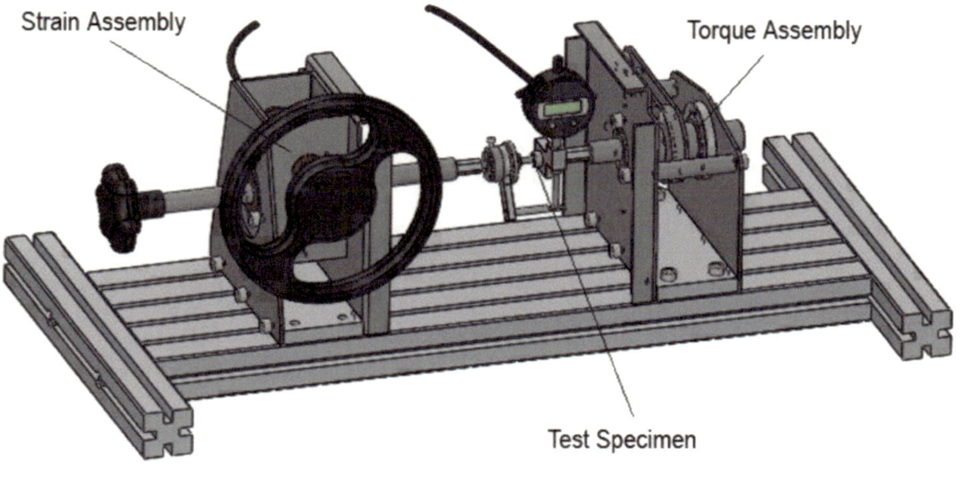

Figure 12.
3D CAD model of a torsion testing machine.

assumptions, the elastic properties of the composite material are fully defined by the Young modulus and shear modulus and the formula involving them allows the calculation of Poisson ratio ν:

$$G = \frac{E}{2(1+\nu)}, \nu = \frac{E}{2G} - 1 \tag{8}$$

The following pages briefly introduce 3D CAD models of tensile testing, torsion, creep and rotating fatigue computer controlled machine recommended in the experimental research of composite materials with natural fibers – see **Figures 11–14**. Screenshots of specific tests are also introduced in **Figures 15–18**.

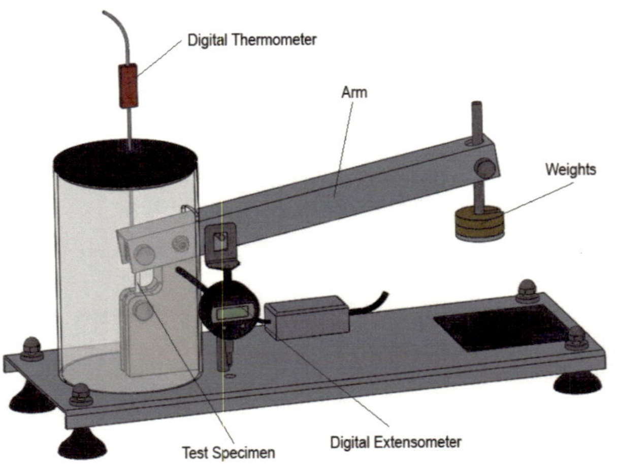

Figure 13.
3D CAD model of a creep testing apparatus.

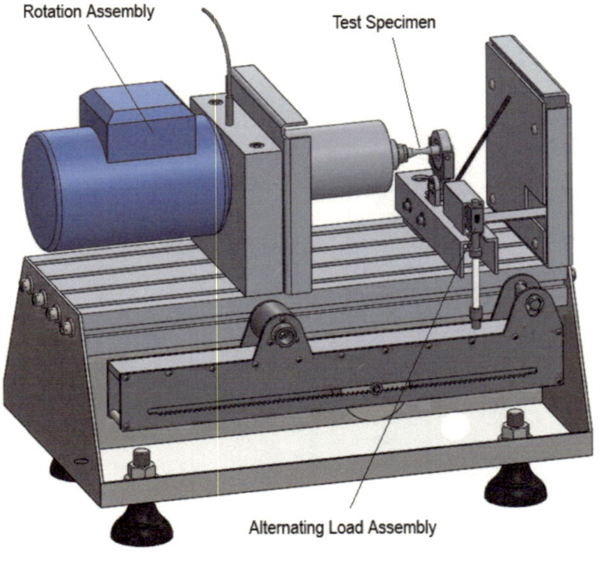

Figure 14.
3D CAD model of a rotating fatigue machine.

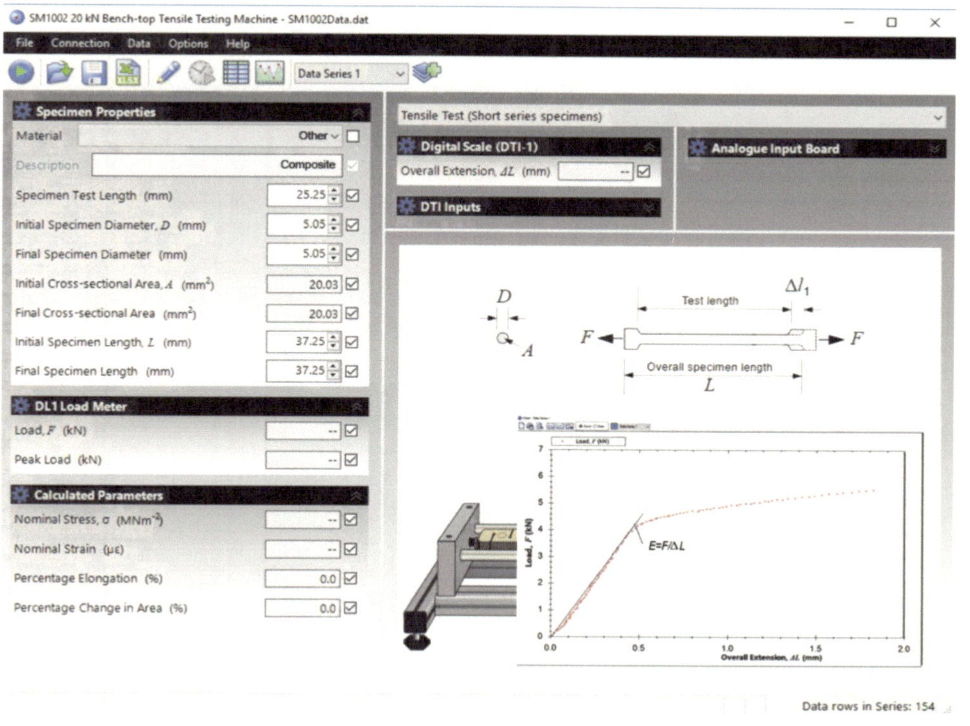

Figure 15.
Tensile test screenshot of a composite material.

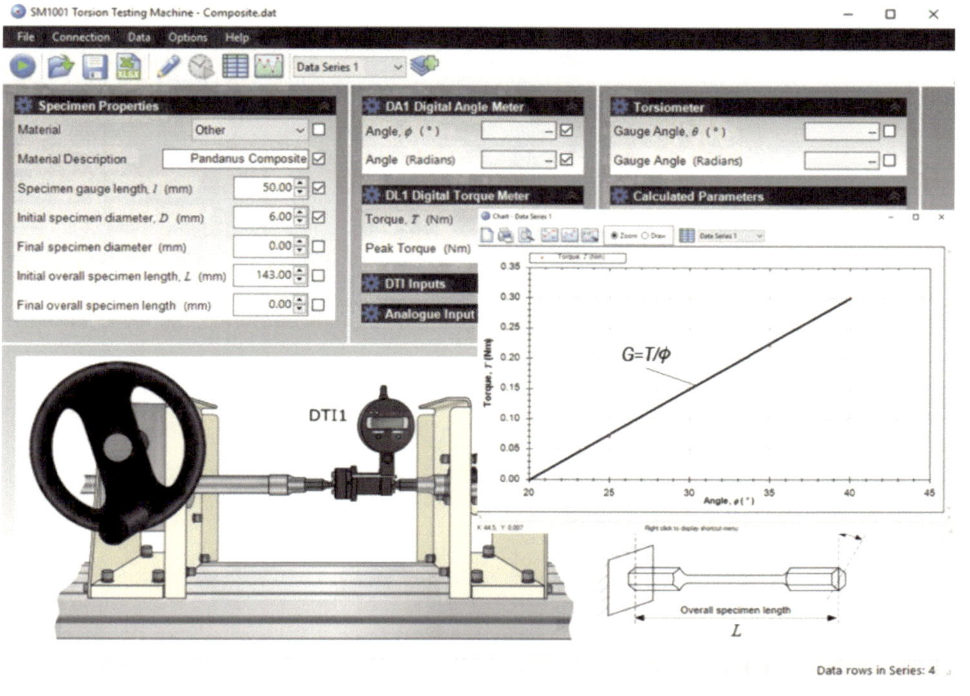

Figure 16.
Torsion test screenshot of a composite material.

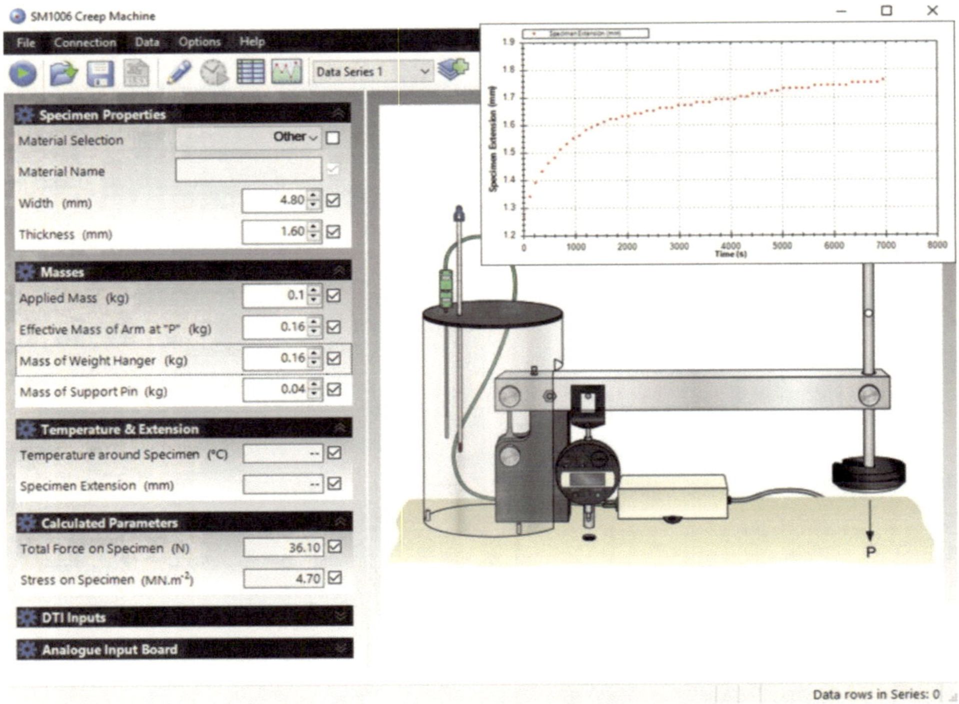

Figure 17.
Creep test screenshot of a composite material.

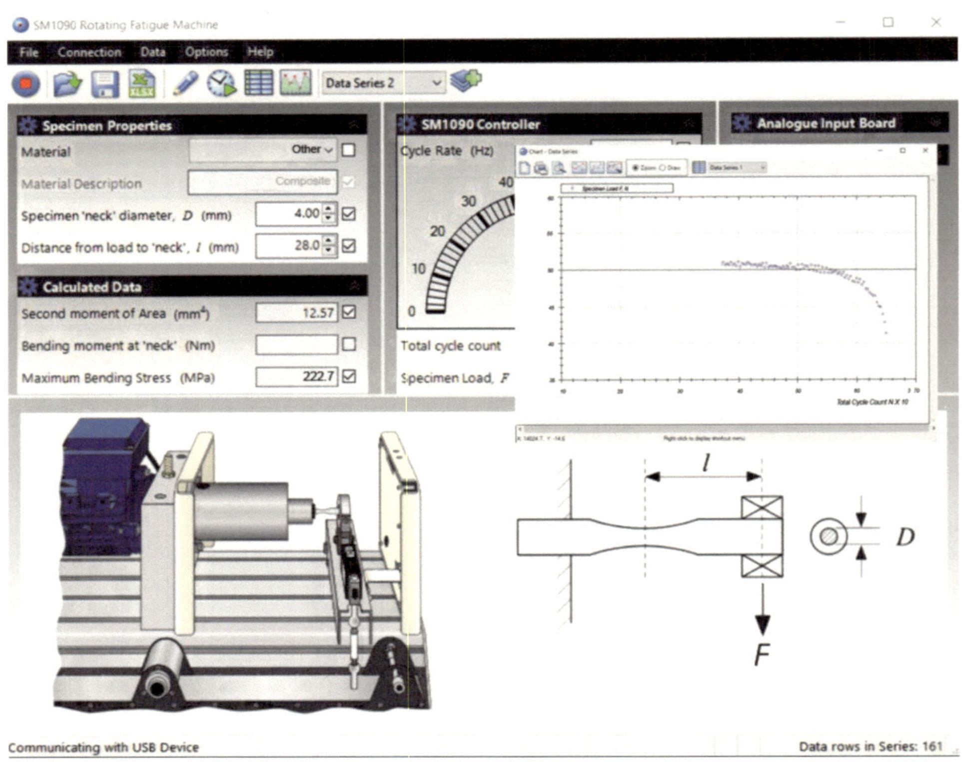

Figure 18.
Fatigue test screenshot of a composite material.

6. Conclusions

Natural fibers show potential as reinforcements in composite materials due to their superior specific strength and their sustainable and recyclable character. Being anisotropic materials, biocomposites are challenging the simulation environments and require advanced and precision experimental equipment. The biohazard aspects of most natural fibers used as reinforcements are minimal. This field of engineering materials is interdisciplinary by its nature and still needs time to mature.

Author details

Nicholas Lambrache[1*], Ora Renagi[1], Lidia Olaru[2] and Brian N'Drelan[1]

1 PNG University of Technology, Lae, Papua New Guinea

2 Jesta Group, Montreal, Canada

*Address all correspondence to: Nicholas.Lambrache@pnguot.ac.pg

References

[1] Luus K. Asbestos: Mining exposure, health effects and policy implications. McGill Journal of Medicine. 2006;**10**(2): 121 6

[2] Fratzl P. Collagen: Structure and Mechanics. New York: Springer. ISBN 978-0-387-73905-2; 2008

[3] Wang B. Keratin: Structure, mechanical properties, occurrence in biological organisms, and efforts at bioinspiration. Progress in Materials Science. 2016;**76**:229-318

[4] Quazi T, Alam K, Quaiyyum M. Mechanical properties of polypropylene composites: A review. Journal of Thermoplastic Composite Materials. 2011;**26**(3):362-391

[5] Fazeli M, Florez J, Simão R. Improvement in adhesion of cellulose fibers to the thermoplastic starch matrix by plasma treatment modification. Composites Part B: Engineering. 2018; **163**:207-216

[6] Hyndman D. Ethnobotany of Wopkaimin Pandanus significant Papua New Guinea plant resource. Economic Botany. 1984;**38**:3

[7] St. John H. Revision of the Genus Pandanus Stickman: New Papuan species in the section microstigma collected by C. E. Carr. Pacific Science. 1968;**22**:4

[8] Thampan PK. Handbook on Coconut Palm. London: Oxford & IBH Publishing; 1981

[9] Bledzki AK, Gassan J. Composites reinforced with cellulose based fibers. Progress in Polymer Science. 1999; **24**(2):221-274

[10] Li X, Tabil TG, Panigrahi S. Chemical treatments of natural fiber for use in natural fiber-reinforced composites: A review. Journal of Polymers and the Environment. 2007; **15**:25-33

[11] Guadalupe M. Preparation and characterization of biodegradable composites based on Brazilian cassava starch, corn starch and green coconut fibers. Carbohydrate Polymers. Vol. 80. Rio de Janeiro, Brazil: Materia. 2010. DOI: 10.1590/S1517-70762010000200034. Available from: https://www.scielo.br/j/rmat/a/MZ3bN4x3thCJm3j8mYck7Km/?lang=en

[12] Awang L. Mechanical properties and microstructure of coconut coir fibers. In Proceeding of MAMIP2012 Asian International Conference on Materials, Minerals and Polymer. 23rd – 24th March 2012, Penang: Research Gate; 2012. Available from: https://www.researchgate.net/publication/236263388_Mechanical_Properties_and_Microstructure_of_Cocos_Nucifera_Coconut_Coir_Fibres

[13] Gentry HS. Agaves of Continental North America. Arizona, United States: University of Arizona Press; 1982. pp. 628-631

[14] Scheller H, Ulvskov P. Hemicelluloses. Annual Revue Plant Biology. 2010;**61**:263-289

[15] Lebo S, Gargulak J, McNally T. Lignin, Kirk-Othmer Encyclopedia of Chemical Technology. New Jersey, United States: John Wiley & Sons; 2001

[16] Lambrache N, Pumwa J, Renagi O, Olaru L, N'Drelan B. Stress behavior of composite materials with natural fibers from the South Pacific. In: Proceedings of the International Conference on Industrial Engineering and Operations Management. Paris, France, United States: IEOM Society International; 2018. pp. 3197-3207

[17] Cook RD. Concepts and Applications of Finite Element Analysis. 2nd ed. New Jersey, United States: John Wiley & Sons; 1981

[18] Von Mises R. Mechanik der festen Körper im plastisch deformablen Zustand. In: Göttingen N. Mathematik Physics. Vol. 1913. France: European Digital Mathematics Library; 1913. pp. 582-592. Available from: https://eudml.org/doc/58894

Chapter 6

Sisal Fibre Based Polymeric Composites

Archana Nigrawal, Arun Kumar Sharma and Fozia Zia Haque

Abstract

Nature origin fibres have drawn in extraordinary consideration from commercial and specialists for the use of polymer composites as a result of their 'greener' nature and commitment to maintainable practice. Different enterprises have moved towards reasonable innovation to work on the harmony between the climate and social and financial concerns. Innovative work has demonstrated that normal fibres have been effectively applied as fortifications in the composites business, for example, for transportation, inside segments, building, airplane.

Keywords: natural fibres, sisal fibres, polymers, composites, reinforcements

1. Introduction

Sisal can be simply cultured in a little cultivated area time. The plant grows in nature in the hedges of fields and railway tracks. Research has showed that approximately 4.5 million tons for every year sisal fibres are extricated all through the planet. Sisal fibre is taken out from trees of the sisal plant (named: Agave sisalana), which is currently planted in tropical parts of the Africa and some other regions of Far East. Usually, sisal plant contains nearly 250 sisal leaves and each leaf is having approximately 1000 to 1200 fibre bundle. On an average, a sisal plant has 4% fibre, 0.75% cuticle, 8% dry substance and 87.25% water. In general, sisal fibre is taken by soaking and by scrapping and using other mechanical resources [1–5].

Higher quantities of supplies and ecological system these days include rise in the claim of industry to use green composite resources. It has also turned out to be the major dynamic compel of current research on the progress of eco-friendly and sustainable natural fibre-toughened polymer composites as an alternative of the synthetic one. An established fact is that that synthetic fibres-reinforced polymer by now has exceptional property and uses. Likewise, glass fibre-based polymeric composites were known for its outstanding properties, which were used in railway track sleepers [1]. Phenol-based glass fibre polymeric composites have high-class fire-resistant property to convene the fire necessity of construction resources [2]. Although from eco-friendly aspect, reinforcement using natural fibres might be a better option as they can be taken out from plant life, birds and farming wastes. From many past years, agriculture-based waste fibres encompass the preferred choice of researchers for its better utilization purposes. Exemplar of agro-waste fibres are palm oil, bagasse fibre, corn fibre, stalks, coir, bamboo, pineapple, banana and rice husk. These fibres are normally extracted from part of the plant such as stem, leaf, seed or even its fruit [3].

It is noteworthy that artificial fibres built up polymer as of now have exceptional properties and uses. It has been reported that glass fibre-supported polymeric composites are having brilliant characteristics and had the option for their use in sleepers of railroad track [1]. Phenol-based glass fibre polymeric composites have great fireproof properties to convene the fire necessity of structure resources [2]. Notwithstanding, the environment-accommodating perspective, support-utilizing normal fibres might be a superior decision as they can be obtained from plants, creatures and agribusiness squanders. Over the previous decade, agribusiness squander fibres have been the most loved options of analysts for its reasonable assets. For example, horticulture squander fibres are oil palm, bagasse, corn, coir, bamboo, pineapple, banana and rice husk. These fibres are ordinarily extricated from parts of the plant such as stem, leaf, seed or even its natural product [3].

Furthermore, natural fibre composites are being less expensive than manufactured composites, which are bio-degradable, richly accessible, sustainable and light in weight. Regular fibres start from three sources, in particular, plants, creatures and minerals. There are more than 2000 kinds of fibre plants on the planet, and these are generally made out of cellulose, for example kenaf, sugar palm, bamboo, corn, cotton, flax, feed (from grass cutting), hemp, henequen, jute, pineapple leaf, banana, ramie and sisal. The utilization of normal fibres in composites can likewise tackle some different issues, for example, moderate energy utilization during production, leaving basically no carbon impression and diminishing removal issues.

The new patterns in the advancement of the more up to date materials have driven in supplanting materials such as glass and carbon built-up composites with the normal fibres-supported composites, for instance in car inside, passer-by connect, transporting beds, composite rooftop tiles, furniture, toys and so on. Notwithstanding, the primary disadvantage of natural fibres as support is that they are incongruent with thermoplastics because of their hydrophilic nature, which brings about the poor interfacial connection between the fibres and grid. This resulted in the poor mechanical properties of the composites. Hence, the change of synthetic fibres is needed to make them less hydrophilic. Endeavours are made to brief about different substance medicines on regular fibres.

Sisal fibre can be essentially refined in a small developed region time. Sisal plant fills within nature inside the fences of fields and railroad track [6–15]. It is removed from the foliage of the sisal plant (known as *A. sisalana*); regularly, any sisal plant contain 200–250 sisal leaves through which each leave can have somewhere around 1000–1200 fibre pack.

Sisal fibre can be taken out from its leaf and can be categorized into three types: mechanical fibre, ribbon fibre and xylem fibre. First type known as mechanical fibres can be taken out from the edge of the leaf, which is similar to horseshoe and can be alienated by the removal process. The second type that is known as ribbon fibre is the greatest fibre and be capable of be tearing up longitudinally throughout its dispensation.

The third type known as xylem fibres are generally uneven in shape and split up easily throughout the dispensation. These fibres crop up in between vascular bundles in contrast to the ribbon fibres [16–19].

In addition to this, the chemical structure of sisal fibre differs from one place to another be dependent on the source of accessibility, measuring techniques, age issue, etc.; similar to additional normal fibres, sisal fibre also contains cellulose, lignin, hemicelluloses and dampness. Sisal fibre contains cellulose 65–68%, hemicellulose 10–22%, lignin 9.9–14% and moisture content 10–22%.

Examination study announced that NFC successfully be utilized in cars, for example, inward motor covers, entryway boards, cap racks, bundle plate, sun visors, seat backs and outside/under floor framing [9], and they have been utilized in inside framing in aviation and airplane enterprises [6, 8–10, 20, 21].

2. Sisal fibre-toughened polymer composites

Along with the different natural fibre-based composites, sisal fibre-supported composite produces predominant effect potency with reasonable ductile and flexural property. It can be used in applications where high-effect strength is required. Presently, sisal fibre utilizes different types of polymers such as thermoset-, thermoplastic- and bio-degradable polymeric-based resources and their different properties have been reported in the literature.

Ramesh et al. [6] researched the mechanical properties of sisal, jute and glass fibre-supported polyester composites and seen that the expansion of glass fibre into jute fibre composite brought about greatest elasticity. Similarly, in the case of jute and sisal combination composites were having maximum flexural force and most extreme effect power was obtained from the sisal fibre composite. Properties of elasticity, flexural strength and compressive strength of epoxy-based sisal-glass-based composites are reported [8, 21] and sisal normal fibre composites are created amid and with no silica by joining fully biodegradable sisal fibres in addition to the polymer framework. The outcomes depicted that the elasticity and pliable modulus of composites with silica are 1.5 and 1.08 occasions more prominent than those of composites without silica individually. The sway strength of composite with sand is 1.36 and 1.8 occasions as compared to the composites without silica what's more, pure polyester, individually. The effect of sisal fibre on the properties of the polymers has been reviewed [9].

Utilizations of sisal fibre and plant fibres show huge commitment in vehicle applications because of its qualities, for example high solidness with light weight per unit region, simple to reuse, 30–40% lighter than glass fibre, decreased fuel utilization, minimal expense, no wear of panels or any part of tiles.

3. Sisal fibre-based polymer composites and their applications

Tooling has no well-being danger, great warm and acoustic protecting properties and so forth. Real modern interest for normal fibres has expanded distinctly in the course of recent years. In 2005, the first run through normal fibres (without wood and cotton) was utilized in car composites [6]. Regular fibre composite materials are being utilized for making a huge number in the car area [21]. Sisal and jute fibres have been utilized in the German auto industry for quite a long time [8]. Mercedes first utilized jute-based entryway boards in quite a while E-Class vehicles in 1996.

As of late, there has been expanding interest in the substitution of glass fibres in built-up plastic composites by normal plant fibres such as flax, hemp and sisal parts [9]. Like glass, the normal fibres consolidate promptly with a thermoplastic or thermosetting grid to deliver item merchandise [10]. The car business requires composite materials to meet execution not really set in stone in a wide scope of tests. Common place market detail incorporates extreme breaking power and stretching, flexural properties, sway strength, hazing qualities, combustibility, acoustic assimilation, appropriateness for preparing: temperature and abide time, scent, water ingestion, dimensional steadiness and crash conduct [11–14, 22]. Plant fibres are as of now just utilized in the inside of traveller vehicles and truck lodges. Other than their utilization in trim parts, for example entryway boards or lodge linings, plant fibres are utilized broadly for thermo-acoustic protection.

Such protecting materials, basically dependent on cotton fibres reused from materials, have moderately high fibre content of over 80% by weight. Trim parts in Brazilian trucks, made of a combination of jute, espresso pack squanders and polypropylene sacks, show that reusing some of the time can prompt progressed applications. Another grounded field of use is the utilization of coconut fibres fortified with regular latex for seat pads. For this application, the capacity of plant fibres to ingest a lot of mugginess prompts an expanded solace that cannot be reached with manufactured materials. Beside this sort of improvements, essentially new applications have not been acknowledged as of late.

Normal fibre composites with thermoplastic and thermoset lattices have been embraced by European vehicle producers and providers for entryway boards, seat backs, main events, bundle plate, dashboards and numerous inside parts. Other natural fibres such as kenaf, hemp, flax, jute and sisal offer such advantages as decrease in weight, cost and CO_2, less reliance on unfamiliar oil sources and recyclability. Glass fibre built-up plastics has demonstrated to meet the underlying and solidness requests of car inside and outside parts. Be that as it may, it displays deficiencies, for example it is somewhat high fibre thickness (40% higher than natural fibres), trouble to machine, helpless reusing property and potential well-being danger. An environmental advancement of normal fibre mat when contrasted with glass fibre mat offers another forthcoming utilization of regular fibre support. Flax, sisal and hemp are prepared into entryway cladding, seatback linings, floor boards and different other car parts [11]. The utilization of plant fibre (sisal/flax/hemp and so forth)-based vehicle parts such as trim parts, different boards, retires and brake shoes are drawing in auto businesses overall on account of its decrease in weight of about 20%, energy creation of 90% and cost of the segments of 15%. Moderate assessments demonstrate that around 7000 TPA plant fibre-based materials can discover their direction keen on traveller vehicles and multi-utility vehicles [9]. Sisal is utilized in entryway cladding, seatback lining and for bundle racks (the gap at the back seats of vehicles).

4. Prospects for use of sisal fibre in automotive manufacturing

Sisal fibres can be utilized in Door boards, Lodge linings; Brake liners; thermo acoustic protection, trim parts, seat pads and back etc. The potential partners for the utility of sisal fibre in auto-segment industry are as follows: Mercedes Benz, Freightliner, Daimler Chrysler, Chevrolet, and General Motors, Mahindra and Mahindra, Tata Motors and Hero Honda and so on. The money-saving advantage investigation, techno-business attainability and the difficulties for sisal fibre double-dealing for different designing applications are as per the following: sisal is xerophytes and fills in badlands, which moderates soil and procures carbon credits. Assured maintainable fibre creation is 2.5 ton/ha for 6–8 years. Surface medicines empower sisal fibres to be utilized as support in a polymer lattice and it has advantageous over mineral and other ordinary regular fibres [10–12].

5. Sisal fibre-based polymer composites and their applications

5.1 Electrical application of sisal fibre

In request to use sisal fibre for electrical applications, a few analysts have considered distinctive electrical properties of sisal fibre at various temperatures and frequencies. Expanding the plant age moves the dissemination factor (tan d) top to higher temperature. Further, the wonders were clarified based on primary charges. Water consumed by sisal fibres has OH anions that go about as dipoles. Other than OH anions, there are a few pollutants and particles on the fibres. At high frequencies, the commitment of polarization of assimilated water particles and space charge diminishes and electronic and nuclear polarization becomes employable. Expansion in temperature influences the portability of particles and subsequently changes the ionic commitments [13, 14]. The electrical properties of sisal fibre built-up LDPE have been concentrated as for the impacts of recurrence, fibre content and fibre length. The dielectric constant increases consistently with expanding fibre concentration for all frequencies in reaching 1–107 Hz. Similarly it is noted that dielectric consistent declines on an increment in fibre length and recurrence. Greatest dielectric consistent qualities are obtained at low frequencies. Sisal/LDPE composites of 1 mm fibre length and 30% fibre concentration contain the most noteworthy upsides of dielectric constants at all frequencies. The upsides of volume receptiveness decline on an increment of recurrence and fibre concentration; that is, the electric conductivity of composites is more prominent than slick LDPE. When contrasted with glass/LDPE composites, similar pattern in electrical properties is noticed; however, the charges of dielectric constants of the last composites on recurrence and fibre concentration are more modest because of their lower interfacial polarization [2–4, 22].

5.2 Application of sisal fibre in railways

Composite materials offer some huge benefits to metals in numerous underlying applications in rail lines such that they are lightweight, practical, consumption safe, energy-lessening underway, and stuff and fire retardant. Composite materials can be utilized in rail routes, the stuff case, primary entryways, gear racks etc.

5.3 Prospects for the use of sisal fibre in construction industries

The addition of fibre strengthening in building materials can improve a lot of the manufacturing properties of the essential resources, such as fracture toughness, flexural strength and resistance to fatigue, impact and thermal shock. In few years, a huge deal of attention has been made all over the worlds on the possible application of natural fibre-based materials and on other construction materials. Research have been done in various nations on different properties such as mechanical physical properties, and toughness of concrete-based matrix toughened with natural fibres such as sisal, coconut, jute, bamboo and wood fibres. Natural fibres are good choice for strengthening of concrete-based materials due to their easy access, less prices as compared with synthetic fibres and less utilization of energy. In this chapter, an effort is made to describe the properties of the natural fibre-based composites [2–4, 22].

6. Conclusion

The inclusion of sisal fibre strengthening in existing technologies can improve a lot of the manufacturing properties of the basic material, such as fracture toughness, flexural strength and resistance to fatigue, impact, thermal shock and spalling.

Acknowledgements

The financial support to Dr. Archana Nigrawal by the DST (DST/WOS-B/AFE-5/2021) project is gratefully acknowledged.

Conflict of interest

The authors declare no conflict of interest.

Author details

Archana Nigrawal[1*], Arun Kumar Sharma[2] and Fozia Zia Haque[1*]

1 Optical Nanomaterials Lab, Department of Physics, Maulana Azad National Institute of Technology, Bhopal, India

2 Department of Mechanical Engineering, Laxmipati Institute of Science and Technology, Bhopal, India

*Address all correspondence to: archananigrawal@yahoo.co.in and foziazia@rediffmail.com

References

[1] Li Y, Mai Y-W, Ye L. Sisal fibre and its composites: A review of recent developments compos. Science and Technology. 2000;**60**:2037-2055

[2] Staiger M, Tucker N. Natural-fibre composites in structural applications Properties and Performance of Natural-Fibre Composites. Elsevier; 2008. pp. 269-300

[3] Mishra S, Mohanty AK, Drzal LT, Misra M, Hinrichsen G. A review on pineapple leaf fibres, sisal fibres and their biocomposites. Macromolecular Materials and Engineering. 2004; **289**:955-974

[4] Ramesh M, Palanikumar K, Reddy KH. Mechanical property evaluation of sisal–jute–glass fibre reinforced polyester composites. Compos. B Eng. 2013;**48**:1-9

[5] Kabir M, Wang H, Lau K, Cardona F. Chemical treatments on plant-based natural fibre reinforced polymer composites: An overview. Compos. B Eng. 2012;**43**:2883-2892

[6] Mohanty M, Misra G, Hinrichsen. Biofibres, biodegradable polymers and biocomposites: An overview. Macromolecular Materials and Engineering. 2000;**276**:1-24

[7] Karus M, Kaup M. Natural fibres in the European automotive industry. Journal of Industrial Hemp. 2002; **7**:119-131

[8] Sathishkumar T, Navaneethakrishnan P, Shankar S, Rajasekar R. Mechanical properties and water absorption of short snake grass fibre reinforced isophthallic polyester composites. Fibres and Polymers. 2014;**15**:1927-1934

[9] Sanjay M, Yogesha B. Studies on natural/glass fibre reinforced polymer hybrid composites: An evolution. Materials Today: Proceedings. 2017; **4**:2739-2747

[10] Yusriah L, Sapuan S, Zainudin E, Mariatti M. Characterization of physical, mechanical, thermal and morphological properties of agro-waste betel nut (Areca catechu) husk fibre. Journal of Cleaner Production. 2014; **72**:174-180

[11] Chandrasekar M, Ishak MR, Sapuan SM, Leman Z, Jawaid M. A review on the characterisation of natural fibres and their composites after alkali treatment and water absorption. Plastics Rubber and Composites. 2017; **46**:119-136

[12] Bessell T, Mutuli S. The interfacial bond strength of sisal—cement composites using a tensile test. Journal of Materials Science Letters. 1982; **1**:244-246

[13] de Andrade Silva F, Toledo Filho RD, de Almeida Melo J, de Almeida Melo Filho J, Fairbairn EMR. Physical and mechanical properties of durable sisal fibre–cement composites. Construction and Building Materials. 2010;**24**:777-785

[14] Savastano H Jr, Warden P, Coutts R. Brazilian waste fibres as reinforcement for cement-based composites. Cement and Concrete Composites. 2000; **22**:379-384

[15] Navin C, Archana N, Deepak J. Dielectric Behaviour of Maleic anhydride grafted polypropylene (MAgPP) Modified Sisal Fibre Reinforced PP Composites. Journal of Natural Fibre. 2008;**5**:270-289

[16] Ramesh M, Palanikumar K, Reddy KH. Mechanical property evaluation of sisal–jute–glass fibre reinforced polyester composites. Composites. Part B, Engineering. 2013;**48**:1-9

[17] Ranganna H et al. Mechanical & thermal properties of epoxy based hybrid composites reinforced with sisal/ glass fibres. International Journal of Fibre and Textile Research. 2012;**2**(3): 26-29

[18] Gowthami et al. Effect of silica on thermal and mechanical properties of sisal fibre rein- forced polyester composites. Environmental Sciences. 2013;**4**(2):199-204

[19] Archana N, Navin C. Preparation, dielectric and A.C. conductivity studies on chemically treated sisal powder filled polyvinyl alcohol (PVA) biocomposites. International Journal of Engineering, Science and Technology. 2012;**4**:119-209

[20] Kabir MM, Wang H, Lau KT, Cardona F. Chemical treatments on plant-based natural fibre reinforced polymer composites: An overview. Composites Part B Engineering. 2012;**43**:2883-2892

[21] Karus M, Kaup M. Natural fibres in the European automotive industry. Journal of Industrial Hemp. 2002; **7**:119-131

[22] Li Y, Mai Y-W, Ye L. Sisal fibre and its composites: A review of recent developments. Composites Science and Technology. 2000;**60**:2037-2055

Chapter 7

Functional Application for the Corn Leaf Fibre to Make Reinforced Polymer Composites Sheet

Ramratan Guru, Anupam Kumar and Rohit Kumar

Abstract

This research work has mainly utilized agricultural waste material to make a good-quality composite sheet product of the profitable, pollution free, economical better for farmer and industries. In this study, from corn leaf fibre to reinforced epoxy composite product has been utilized with minimum 35 to maximum range 55% but according to earlier studies, pulp composite material was used in minimum 10 to maximum 27%. Natural fibre-based composites are under intensive study due to their light weight, eco-friendly nature and unique properties. Due to the continuous supply, easy of handling, safety and biodegradability, natural fibre is considered as better alternative in replacing many structural and non-structural components. Corn leaf fibre pulp can be new source of raw material to the industries and can be potential replacement for the expensive and non-renewable synthetic fibre. Corn leaf fibre as the filler material and epoxy as the matrix material were used by changing reinforcement weight fraction. Composites were prepared using hand lay-up techniques by maintaining constant fibre and matrix volume fraction. The sample of the composites thus fabricated was subjected to tensile, impact test for finding the effect of corn husk in different concentrations.

Keywords: corn leaf fibre, tensile strength testing, impact strength testing, epoxy resin

1. Introduction

Nowadays, most developed countries are paying special attention to environmental issues, and some of the most important actions to protect the environment are focused on the optimum use of natural resource, the reduction of air pollution, loss of soil moisture, upgrading industrial and agroforestry wastes, etc. [1–3]. All these are with main aim of reducing and utilizing agricultural waste, which may be profitable, pollution free and economically viable for the farmer and industries. The most suitable way is to solve environmental related issues and increase the income of farmers [4–6]. Corn leaf fibre is nowadays exploited as reinforcements material owing to their low cost, fairly good mechanical properties, high specific strength, availability, eco-friendly and biodegradability characteristics

Application of high-performance composites using natural fibres is increasing in various engineering fields. This not only reduces the cost but also saves from environmental pollution [7–10]. These composites are also used in panel for partition and false ceiling, wall sheet, floor, window and door frame. The mainly fibre-reinforced composite is handled to have material properties in all directions and material properties are different in all performances with control of the fibre orientation [11–14].

2. Materials and methods

2.1 Materials

The corn leaf agricultural waste material was collected from local corn farm. The chemical constituent of corn leaves was found to be cellulose of 43%, lignin of 22%, hemicelluloses of 31% and Ash of 1.9% (**Figure 1**).

2.2 Preparation of corn leaf fibre

The corn leaves are cut into 5 to 6 cm length and 1 kg dried leaves are placed in the cooking pot. Make a solution of sodium hydroxide and soda AQ, pour the solution in the pot with leaves and boil for 2 to 3 hour After that, it will become pulp and we will wait for it to cool down and wash it with water three or four times, and then, we will blend with the help of blending machine and squeeze them the help of cotton fabric (**Table 1**) [15, 16].

2.3 Preparation of corn leaf fibre polyester composites

The polyester resin with hardener is mixed in a container and then continuously stirred for 2 to 5 minutes. Then, corn leaf fibre was added with continuous stirring and uniform mixture. Afterwards, this prepared liquid material is placed on the mould, and after 3 hour, mould is removed as composite sheet. This final composite sheet is properly placed in contact with sunlight for 5 to 7 days (**Figure 2**).

Figure 1.
Corn leaf.

S. No	Properties	Range (%)
1.	Cellulose	43
2.	Lignin	22
3.	Hemicelluloses	31
4.	Ash	1.9

Table 1.
Chemical composition of corn leaf.

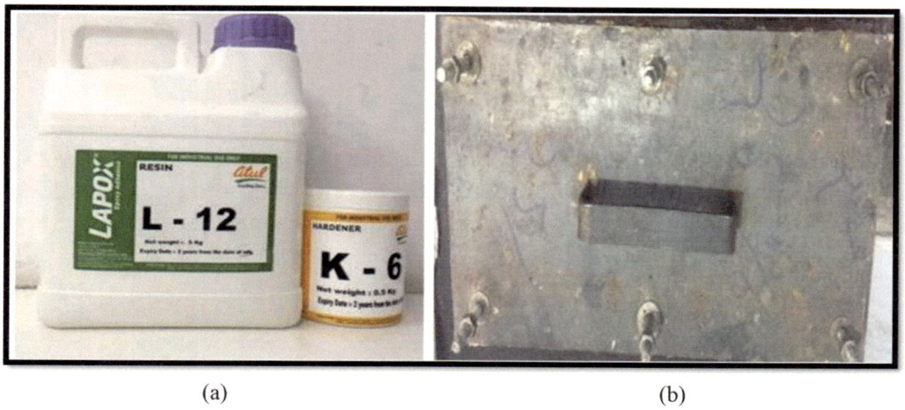

Figure 2.
(a) Resin and hardener. (b) Mould.

Figure 3.
All natural fibre images.

S. No	Thickness(mm)	Pulp ratio (%)	Epoxy (%)	Hardener (%)
1.	4	35	50	15
2.	4	45	40	15
3.	4	55	30	15
4.	8	35	50	15
5.	8	45	40	15
6.	8	55	30	15

Table 2.
Specimen composition.

2.4 Reason to choose corn leaf

There are severe environmental problems such as air pollution, smoke formation and loss of soil moisture. The main aim to choose corn leaf is utilization of agricultural waste that may be profitably pollution free and economically viable for the farmer (**Figure 3** and **Table 2**).

3. Methods

3.1 Thickness test

Thickness measurement instruments are used as per standard performance ASTM D-1777.

3.2 Tensile strength testing

Tensile measurement instrument is used as per ASTMD--638 standards. This instrument is of very good quality and has properly accurate measurement material sheet with tensile result and easily operating system (**Figures 4** and **5**, **Table 3**).

Tensile strength (TS) = load maximum/specimen of the material.

3.3 Impact strength

The specimen was exposed to a large amount of power for a small intermission of stage. A measurable with more impact energy will have more roughness. This impact measurement instrument is used for as per ASTMD-256 standards (**Figures 6** and 7; **Tables 4–6**).

4. Results and discussions

Mechanical properties of corn leaf fibre and epoxy resin composites are explained here.

4.1 Density for the composite sheet

It can be observed from **Table** 7 and **Figure 8** that with pulp ratio (35%) and thickness of 4 mm and 8 mm, density values are 1.183 g/m^3 and 0.987 g/m^3,

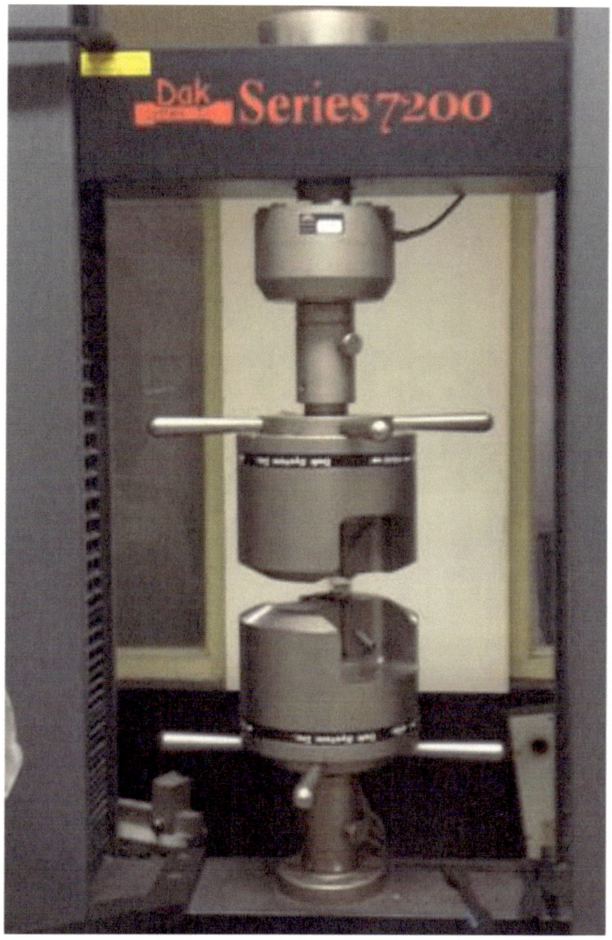

Figure 4.
Tensile strength testing instrument.

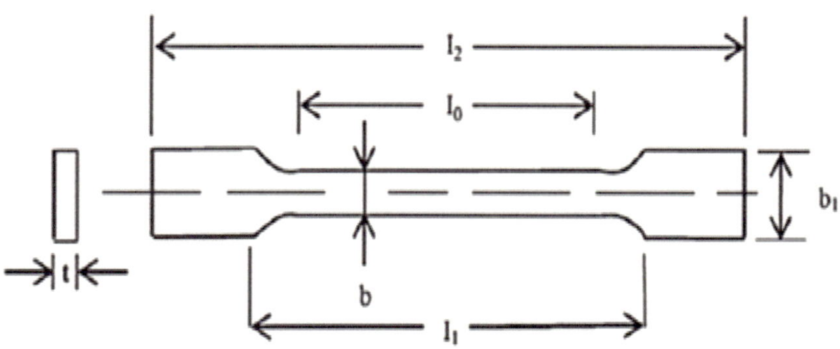

Figure 5.
Tensile specimen shape.

respectively. Further, it is observed that at low pulp ratio (%), density (g/m^3) for composite sheet materials is higher. With pulp ratio of 55% in composite sheet and for both thickness levels 4 mm and 8 mm, lower density is observed. Increase in pulp ratio % in sample results in decrease in density for composite material sheet.

Sr.no.	Symbol	Description	Dimension (mm)
1.	I0	Gauge length	145
2.	I1	Grip distance	160
3.	I2	Overall length	240
4.	b	Width of narrow parallel portion	15
5.	b1	Width ends	30
6.	T	Thickness	4, 8

Table 3.
Tensile specimen dimensions.

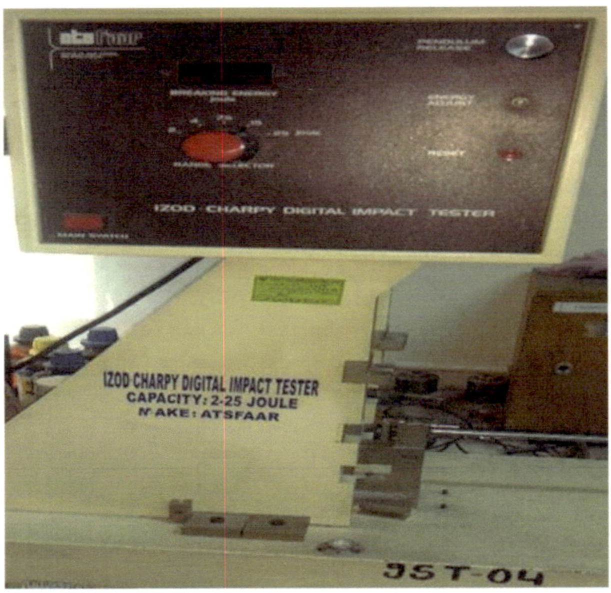

Figure 6.
Impact strength testing instruments.

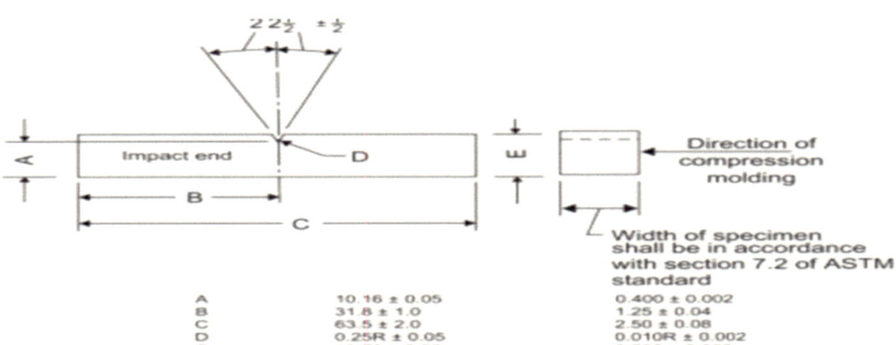

Figure 7.
Impact specimen of IZOD shape.

Sr. no.	Description	Dimension (mm)
1.	Length of specimen	63.5
2.	Centring of notch	2.54
3.	Radius of notch type V	0.25
4.	Angle of notch	45′
5.	Width	12.70
6.	Thickness	4, 8

Table 4.
Impact specimen dimensions.

Sr.no.	Pulp ratio (%)	Various thickness (mm)	Mean thickness (mm)
1.	35	4.16, 4.22, 4.28, 4.26, 4.33, 4.48, 4.56, 4.87, 4.10, 4.21	4.347
2.	45	4.33, 4.42, 4.37, 4.56, 4.68, 4.77, 4.87, 4.66, 5.02, 5.13	4.681
3.	55	4.56, 4.68, 4.72, 4.26, 4.32, 4.37, 4.76, 5.12, 5.33, 5.22	4.734
4.	35	7.66, 7.90, 7.88, 7.97, 8.10, 8.34, 8.44, 8.56, 8.69, 8.78	8.232
5.	45	8.24, 8.33, 8.54, 8.56, 8.76, 8.87, 8.42, 8.19, 9.09, 9.18	9.681
6.	55	8.16, 8.23, 8.36, 8.40, 8.27, 8.26, 8.41, 8.43, 8.53, 8.66	8.371

Table 5.
Sheet thickness results.

Sr.no.	Thickness(mm)	Pulp ratio (%)	Weight (gm)	GSM(g/m^2)
1.	4	35	142	142gm/0.03 = 4733
2.	4	45	136	136gm/0.03 = 4533
3.	4	55	124	124gm/0.03 = 4133
4.	8	35	237	237gm/0.03 = 7900
5.	8	45	223	223gm/0.03 = 7433
6.	8	55	210	210gm/0.03 = 7000

Table 6.
Sample GSM.

Sr.no.	Thickness (mm)	Pulp ratio (%)	Density (g/m^3)
1.	4	35	142gm/120 = 1.1833
2.	4	45	136gm/120 = 1.1333
3.	4	55	124gm/120 = 1.0333
4.	8	35	237gm/240 = 0.9875
5.	8	45	223gm/240 = 0.9291
6.	8	55	210gm/240 = 0.8750

Table 7.
Sample density.

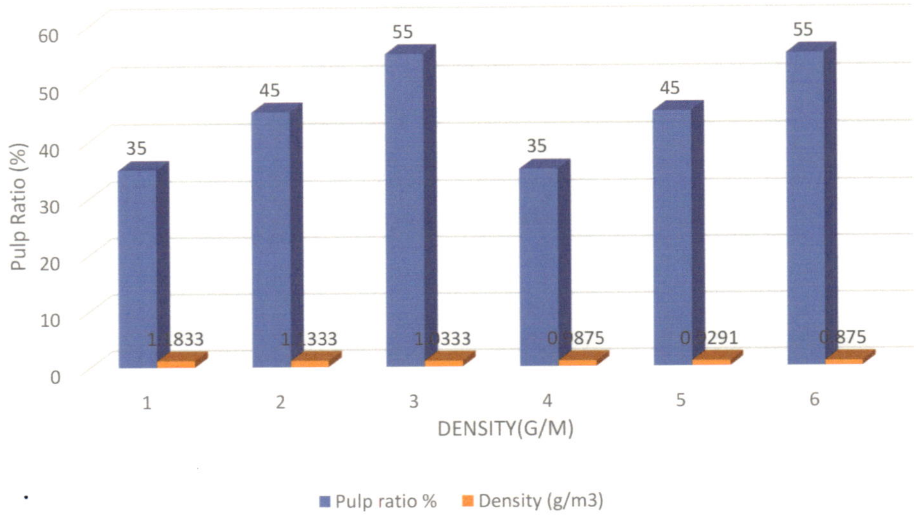

Figure 8.
Effect of the pulp ratio and thickness on density properties.

Figure 9.
Image for composite sheet material.

Figure 10.
Specimen size for composite sheet material.

This may be due to the presence of some voids. It can be seen from the images that void is there between fibre and matrix, which shows less compatibility of corn leaf fibre with epoxy resin at higher pulp ratio (**Figures 9–11**).

Figure 11.
Specimen break after tensile strength test for composite material.

4.2 Tensile strength properties for composite sheet material

Samples are tested on DAK series 7200 machine (a universal testing machine) and tensile strength is recorded. The values are depicted in **Table 8**.

It can be observed from **Table 8** and **Figure 12** that at pulp ratio of 35%, a thickness level of 4 mm and 8 mm gives the highest tensile strength value 4.95 Mpa and 7.58 Mpa, respectively. It may be due to the higher density for composite sheet according to **Table 7**. The maximum tensile strength obtained for corn leaf fibre ratio (35%) may also be due to good bonding between epoxy and fibre pulp ratio, and approximately hexagonal structure of cellulose present in the sugarcane leaves fibre gives high flexibility to the fibre. Further, higher tensile strength may be attributed to the more energy transfer to the fibres by the matrix [17, 18].

With pulp ratio (55%) in composite sheet and for both 4 mm and 8 mm thickness, low tensile strength has been observed. It may be due to increase in pulp ratio % decreases density as reported earlier. The decrease in tensile strength with increase in pulp ratio (%) may be due to increase in voids created by air entrapment due to small-size particles. The tensile strength for neat epoxy is very less, which shows that there is decrease in strength in composite and less density that may be due to weak adhesion between fibre pulp and matrix, which causes no proper transfer of energy to the fibres by matrix.

4.3 Impact strength properties for composite sheet material

In order to find the impact capability of sample, impact test is carried out using IZOD – CHARPY digital impact tester machine. Absorbed energy obtained for six

Sr.no.	Pulp ratio (%)	Thickness (mm)	Tensile strength 1 (Mpa)ASTMD-638	Tensile strength 2(Mpa)ASTMD-638	Mean tensile strength
1.	35	4	5.47	4.43	4.950
2.	45	4	2.10	4.17	3.135
3.	55	4	1.98	3.15	2.565
4.	35	8	7.61	7.58	7.595
5.	45	8	4.47	5.10	4.785
6.	55	8	3.88	4.02	3.950

Table 8.
Tensile strength result.

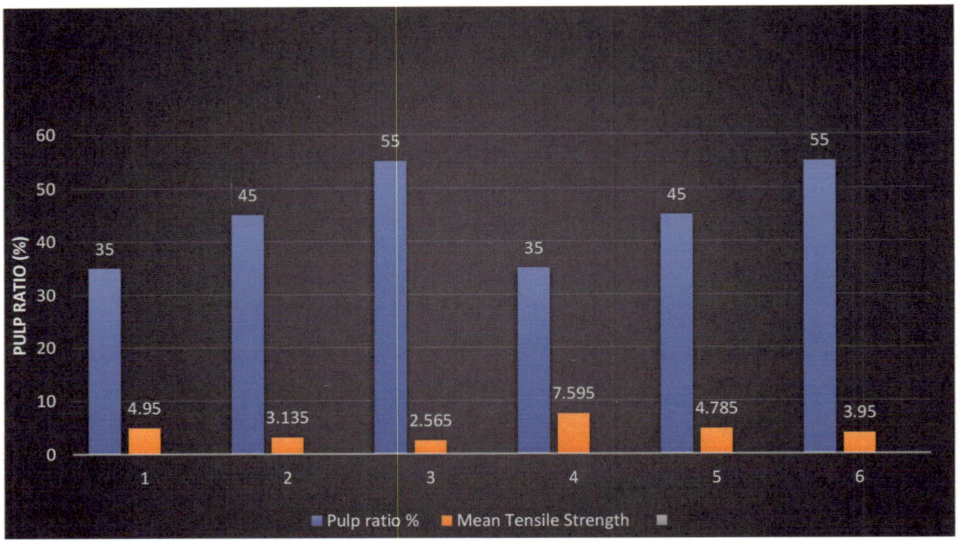

Figure 12.
Effect of the pulp ratio and thickness on tensile properties.

Sr.no.	Pulp ratio%	Thickness (mm)	Impact strength 1(kg. / m2)ASTMD-256	Impact strength 2(kg. /m^2) ASTMD-256	Impact strength
1.	35	4	1.77	1.35	1.11
2.	45	4	1.25	1.49	1.55
3.	55	4	1.22	1.37	1.79
4.	35	8	2.39	2.10	1.89
5.	45	8	1.35	1.67	1.99
6.	55	8	2.43	1.72	2.40

Table 9.
Impact test results.

different (samples) composite sheets from the IZOD – CHARPY machine has been shown in figure. It has been observed from **Table 9** that maximum impact strength of 2.40 Kj/m^2 is obtained for 55% pulp ratio sheet.

Further, pulp ratio of (55%) with 4 mm and 8 mm thickness has highest impact strength for 1.79 Kj/m^2 and 2.4 Kj/m^2 respectively. The impact strength of the composites increases with increase in fibre content (pulp ratio %). It may be due to during impact force, fibre pulls out and fibre breakage occurs at lower fibre loading of the polymer composites. At high fibre loading, the fibre crowding leads to easy debonding, which in turn increases the impact resistance, when compared with the impact energy of neat epoxy which is very less. It is observed that the impact strength increases.

Pulp ratio (35%) in composite sheet for both thickness levels of 4 and 8 mm, lower impact strength has been observed with decrease in pulp ratio % and epoxy resin concentration level. The optimum combination for the impact strength of the composites is pulp ratio of 55% and for both thickness levels as seen in **Figures 13** and **14**, respectively.

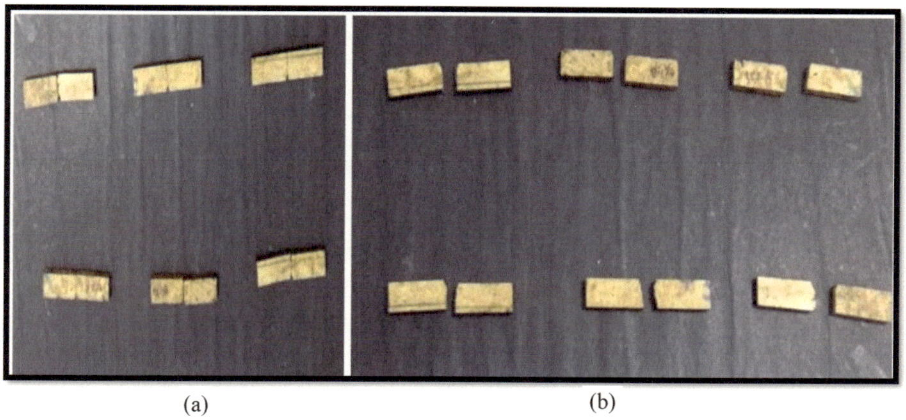

Figure 13.
(a). Impact specimen (b). Broken specimen after impact strength testing.

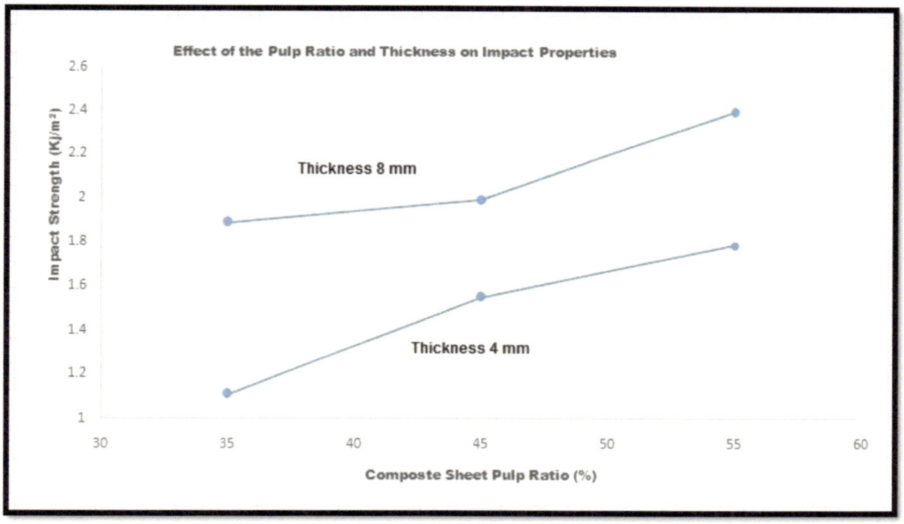

Figure 14.
Effect of the pulp ratio and thickness on impact strength properties.

4.4 Fractured surface morphology using field emission scanning electron microscope (FESEM)

Scanning electron microscope (SEM) is basically measurement the material surface by scanning through high wavelength beams with different wavelengths. The different composite materials are shown in SEM image **Figures 15–17**.

Field emission scanning electron microscope image for sugarcane leaf fibre pulp-reinforced composites is taken to understand microstructure. From the images, the fibre dispersion, matrix without fibres and fibre agglomeration are clearly visible. The images indicate that the failure is due to matrix failure, fibre pull-out and also due to the presence of some voids. The FESEM image of coir fibre-reinforced samples clearly shows the broken sugarcane leaf fibres. It is seen

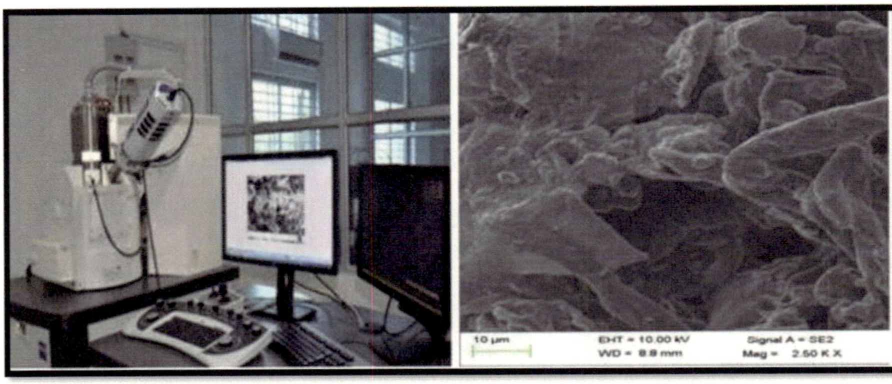

Figure 15.
FESEM image for pulp ratio and epoxy resin for composite sheet.

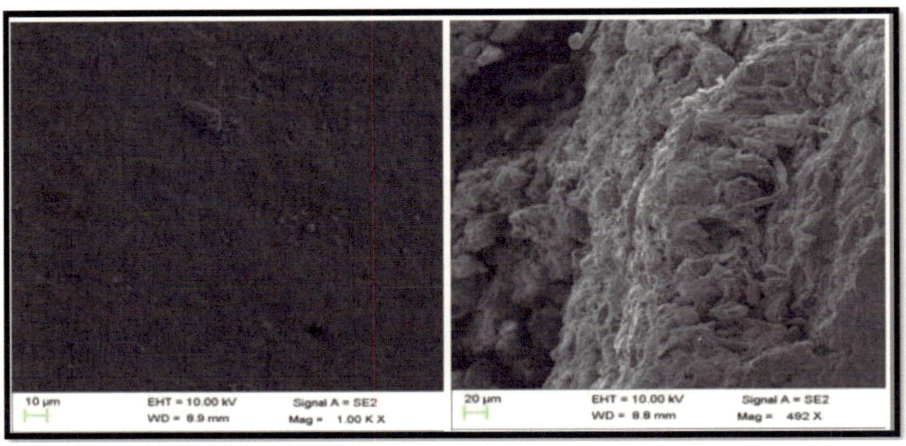

Figure 16.
FESEM image for pulp ratio and epoxy resin for composite sheet.

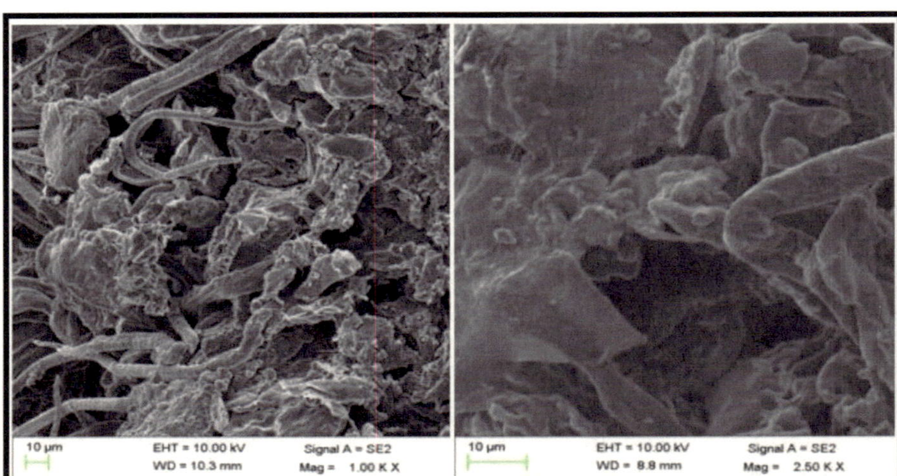

Figure 17.
FESEM image for pulp ratio and epoxy resin for composite sheet.

from the images that void is there between fibre and matrix, which shows less compatibility of corn leaf fibre with epoxy resin. Uniform mixing are pulp ratios and other particles leads to less air voids.

5. Conclusion

This study on the natural corn leaf fibre-reinforced epoxy composites uses pulp ratio (%), thickness level of 4 mm and 8 mm and constant hardness as filler material for the following conclusions: From the present work, it is concluded that the natural corn leaf fibre along with pulp ratio (%), thickness level of 4 mm and 8 mm and constant hardness can be successfully used with epoxy resin to prepare polymer matrix composites with good mechanical properties.

- For composite sheet, it is found that pulp ratio is of 35% and thickness of 4 mm to highest tensile strength 4.95. According to figure, it is found that for pulp ratio of 35% and thickness of 8 mm to the highest tensile strength 7.59. It is observed that the maximum value of tensile strength is obtained for corn leaf fibre ratio (35%) composite sheet. Increasing pulp ration (%) and low density reduce tensile strength due to non-bonding of pulp particles with sugarcane leaf fibre.

- It is found that pulp ratio is of 55% and thickness of 4 mm to highest impact strength for about 1.79. According to figure, it is found that pulp ratio of 55% and thickness of 8 mm give the highest impact strength 2.4. The impact strength of the composites increases with increase in fibre content (pulp ratio %), and this is because on the application of the impact force, fibre pulls out and fibre breakage occurs at lower fibre loading of the polymer composites. The pulp ratio (35%) in composite sheet and both levels of thickness have lower impact strength.

- The FESEM image shows good bonding between fibre, pulp ration (%) and matrix. Also, breakage of fibres shows the load transferred from matrix to the fibres.

- The use of natural sugarcane leaf fibres is to prepare eco-friendly composites. It is utilized as agricultural waste, which may be profitable, pollution free and economically viable for the farmer and industries. These composites are also used in panel for partition and false ceiling, wall sheet, floor, window and door frame.

Author details

Ramratan Guru[1]*, Anupam Kumar[2] and Rohit Kumar[2]

1 Department of Handloom and Textile Technology, Indian Institute of Handloom Technology, Varanasi, India

2 Department of Textile Engineering, Giani Zail Singh Campus College of Engineering and Technology, MRSPTU, Bathinda, Punjab, India

*Address all correspondence to: ramratan333@gmail.com

References

[1] Youssef AM, El-Gendy A, Kamel S. Evaluation of corn husk fibers reinforced recycled low-density polyethylene composites. Materials Chemistry and Physics. 2015;**152**(1): 26-33

[2] Shah H, Yang Y. Chemically extracted corn-husk fibers as reinforcement in light weight poly (propylene) composites. Macromolecular Materials and Engineering. 2008;**29393**:235-243

[3] Majhi S, Samantarai SP, Acharya SK. Tribological behavior of modified rice husk filled epoxy composite. International Journal of Scientific and Engineering Research. 2012;**3**(6):1-5

[4] Nazrul IM. Physico-mechanical properties of chemically treated coir reinforced polypropylene composites. Composites Part A: Applied Science and Manufacturing. 2010;**41**(2):192-198

[5] Jawaid MHPS, Abdul HPS, Khalil. Cellulosic/synthetic fiber reinforced polymer hybrid composites: A review. Carbohydrate Polymers. 2011;**86**(1):1-18

[6] Nuruzzaman DM, Chowdhury MA. Friction and Wear of Polymer and Composites. Rijeka: INTECH Open Access Publisher; 2012

[7] Shalwan A, Yousif BF. In state of art: Mechanical and tribological behavior of polymeric composites based on natural fibers. Materials & Design. 2013;**48**(1): 14-24

[8] Isa MT. The Effect of fiber treatment on the mechanical and water absorption properties of short Okra/Glass fibers hybridized epoxy composites. Inter-National Journal of Materials Engineering. 2014;**4**(5):180-184

[9] Norashikin MZ, Ibrahim MZ. The potential of natural waste (corn husk) for production of environmentally friendly biodegradable film for seedling. World Academy of Science, Engineering and Technology. 2009;**58**(1):176-180

[10] Menezes PL, Rohatgi PK, Lovell MR. Studies on the tribological behavior of natural fiber reinforced polymer composite. In: Green Tribology. Vol. 2(1). Springer Berlin Heidelberg; 2012. pp. 329-345

[11] Aireddy H, Mishra SC. Tribological behavior and mechanical properties of Bio-waste reinforced polymer matrix composites. Journal of Metallurgy and Materials Science. 2011;**5**(2):139-152

[12] Yousif BF, Nirmal U, Wong KJ. Three body abrasion on wear and frictional performance of treated betelnut fiber reinforced epoxy (T-BFRE) composite. Materials & Design. 2010;**31**(9):4514-4521

[13] Pihtili H. An experimental investigation of wear of glass fiber epoxy resin and glass fiber–polyester resin composite materials. European Polymer Journal. 2009;**45**(1):149-154

[14] Brahim SB, Cheikh RB. Influence of fiber orientation and volume fraction on the tensile properties of unidirectional Alfa-polyester composite. Composites Science and Technology. 2006;**67**(1): 140-147

[15] Brahmakumar M, Pavithran C, Pillai RM. Coconut fiber reinforced polyethylene composite: Effect of natural waxy surface layer of the Fibercon fiber/ Matrix interfacial bonding and strength of composites. Composites Science and Technology. 2005;**65**(2):563-569

[16] Roe P, Ansell MP. Jute-reinforced polyester composites. Journal of Materials Science. 1985;**20**(11):4015-4020

[17] Shah DU. Characterisation and Optimisation of the Mechanical

Performance of Plant Fibre Composites for Structural Applications. United Kingdom: University of Nottingham; 2013. p. 274

[18] Mutasher SA, Poh A, Than AM, Law J. The effect of alkali treatment mechanical properties of kenaf fiber epoxy composite. In: Engineering Materials. Trans Tech Publications; 2011

Section 4

FRP Application

Chapter 8

Fabricating Natural Biocomposites for Food Packaging

Liqaa Hamid and Irene Samy

Abstract

Nowadays, there are dominant scientific breakthroughs to advance the packaging industry to identify innovative and emerging fruitful results for making the food packaging systems, in particular, more efficient, resilient, and sustainable. Therefore, friendliness packaging research has been gaining momentum, thanks to global environmental awareness, and also consumer ecological consciousness, and leading companies are committing to a more holistic worldview of packaging in response to more sustainable processes to reduce pollution and any depletion of resources. High-yielding and cost-effective production and design of packaging, involving synthetic materials use reduction and development of new bio-based packaging materials, are very much part of this holistic approach. Thus, in comparison with petroleum-based materials, potential bio-based materials may have benefits for all agents comprised: the producers, customers as well as the whole environment. This chapter explores a review of relative topics across all disciplines that could accelerate understanding toward this goal. It walks through conventional materials, and then important natural and synthetic polymers from the context of food packaging. Moreover, it provides an overview of the performance of bioplastics and their limitations. State-of-the-art main trends on green biocomposites thereof, their potential to transform the food industry, are also herein considered.

Keywords: bioplastics, food packaging, sustainable, polymers, biocomposites

1. Introduction

Food packaging is basically defined as the process of enclosing food to keep it from spoiling by any probable contamination sources [1]. It is at the heart of the current food industry because almost no foods are sold unpackaged. Food packaging is an important step in the long journey starting from farmer to customer, including packaging that must be user-friendly for handling, transporting, and marketing produce ending with their appeal to the consumer [2]. Packaging and its materials add a major expense to the produce industry; therefore, it is critical that packers, shippers, buyers, and consumers understand the full range of available packaging options [3]. Packaging has made it possible to have year-round access to many foods that would not have been possible otherwise. Food packaging has evolved from being simply a container for food to something that can play an important role in food quality by having various characteristics that have been developed to protect the food. Good packaging reduces waste and ensures that the food retains its desired quality for the meant duration of its shelf life.

Today, there are lots of packaging materials and designs on our sides. It should be noted that some packages have actually allowed for the creation of new supermarket categories. Examples include fresh-cut produce and microwave popcorn bags, which owe their existence to the suitable packaging that has been figured out just right [4]. Despite its importance and the vital role it plays, the packaging is often considered as to some extent unnecessary, a serious waste of resources, and an environmental menace. Such points of view arise as a result of the fact that, by the time most consumers come into contact with a package, its job has, in many cases, been done. Consumer demand, on the other hand, will change for years to come as the quality and quantity of food packaging must significantly increase [5]. The industry of food packaging is mainly depending on the utilization of polyolefins (POs). Polyolefin is a comprehensive term for the two mostly used plastics in food packaging, polyethylene (PE), and polypropylene (PP) [6]. Both have an effective combination of properties such as strength, lightness, flexibility, stability, and ease of processing. However, whether they are suitable for recycling and reuse is debatable. At the end of the twentieth century, plastics were found to be persistent pollutants in many environmental niches since they are massively nonbiodegradable. Accordingly, environmental pollution led to an increased interest in developing more sustainable materials in order to deal with the negative effects of plastic pollution [7]. It is important to recognize the potential reduction in pollution that would result from reusing and recycling materials. Hence, the world begins to initiate movements that will clearly eliminate the unfriendly environmental impacts resulted from the existing packaging systems. A new system is required in which consumption is reduced and materials are recycled. A solution can be found in biologically augmented materials. Nowadays, the concerned societies harbor the utilization of natural-based materials more and often. The main attempt resembling that is the turning over from plastics to natural fiber composites (NFCs) as a more so of applicable green alternatives [8].

2. Food packaging science and emerging trends

Food packaging incorporates a wide range of technological activities, including design, package fabrication, shelf-life testing, distribution, and marketing. Food packaging science is a discipline that applies the perfect blend of four pioneer science areas: materials science, food science, information science, and socio-economics in order to understand the requirements of any packaging system [9]. Material science and food science have been the two major guiding principles in the development of food packaging. Material science is essential to comprehend the appearance, mechanical strength, barrier, and physical and chemical properties of any substance used [10]. Food science is concerned with the kinetics of food deterioration and governing its shelf-life. Food packaging technology is a science-based solution to specific food packaging needs; examples include tamper-proof packaging, microwavable packaging, and modified atmospheric packaging, all of which aim to improve consumer safety, quality, and convenience [11]. The primary goal of the field of information science is to examine consumer attitudes and behaviors toward food packaging, and then analyze the results to determine what the most important features of packaging are for consumers, which are comfort for use and durability [12]. In regard to socioeconomic needs, the very distinguished ones are lifestyle changes, in a way that reflects the quality of life through the added value offered to consumers through packaged products, profitable companies, packaged product safety, and environmental protection [13].

2.1 Traditional vs. modern packaging

Traditional or conventional packaging deals with the utilization of wooden or paper parts for packaging perishable food products such as vegetables and fruits. This kind of classical packaging as an option often fails due to several whys, including, it is only viable, therefore, for those products that are showing natural protection against influencing elements resembling daylight, moisture, pathogens, and contaminants [14]. Moreover, many use a fragile material that can be broken with the blows, hence inefficiency, and some materials are heavy, therefore incompatible to storage and transport easily. Presently, these methods are being avoided and hardly practiced anywhere. This led to the development of a new packaging option for food products [15].

Food product development innovations and packaging technologies are the lifeblood of the food industry. Currently, the scope is to meet consumers' demand for nutritious and safe food while minimizing the negative environmental effects of food packaging [16]. This has posed a challenge to replace the existing food packaging practices and drove the emergence of modern packaging approaches. Modern technologies, as opposed to the traditional mode of packaging, are a current necessity. These technologies are related to tailored and elegant packaging, but they also ensure preservation, increased shelf life, the wholesomeness of the product, and ease of transportation and storage. Such technologies can be used for various personalized applications, for instance, extension of shelf life and eliminating risks of contamination. Modern packaging technologies have already transformed the accessibility of a product. Novel packaging technology enabled the adjustment of preservation formula to food products of varying origins, physicochemical characteristics, and sizes without losing their nutritional value [17]. Latest packaging materials and technologies involve green innovations including biopolymers and smart packaging, just to name a few. In the case of food packaging, smart packaging is highly regarded because it provides numerous benefits, including further functional properties similar to antioxidant and antibacterial. Such effort may prolong the product's shelf life and reduce food spoilage and loss. These state-of-the-art packaging technologies in synergy with innovations in food product designs will provide new plentiful pioneer opportunities for the food industry [18]. These state-of-the-art packaging technologies, when combined with innovations in food product design, will create a plethora of new groundbreaking opportunities for the food industry.

3. Food packaging materials

The aim when the food is packaged mostly is to be safely transported and correctly stored. The proper combination of packaging materials and technologies ensures that product quality and freshness are maintained during distribution and storage. Package design and structure are important considerations. That is, not only a container, but it must also protect what it sells and sell what it protects [9]. One of the most important aspects of packaging from a business standpoint is that it identifies the appearance of the product in the distribution chain and distinguishes it when reaching the consumer. Today's food packaging frequently combines several materials to take advantage of each material's functional or esthetic properties. Advances in the field of food packaging research may have an impact on the environment as it continues. Aside from food preservation, current techniques are focused on two extra goals: the suitability of the materials used and the generation of eco-friendly products with no presence of side effects on health.

Food packaging materials that have traditionally been used include wood, paper, and polymers, to most new biocomposites. Plastics do not fare well, but neither do wood or paper. However, there is a shift in demand, which makes biocomposites the finest among all [19].

3.1 Wood

Wood has naturally favorable mechanical, physical, and chemical properties, which makes it an excellent packaging material that possesses a good weight-to-strength ratio, and on the other hand, has a high hydration capacity. Aside from its natural moisture content, wood can absorb or release water in equilibrium with the food with which it comes into direct contact. This is one of the characteristics that, for example, seafood, vegetable, and fruit industries stakeholders seek [20]. Furthermore, the majority of wood species have the naturally occurring acidic pH range of 4.3–5.2, which influences bacterial survival on their surfaces [21]. Wooden materials are less expensive to produce than other containers and are still widely used. Depending on the country's regulations and marketing chains, they can be considered for both single-use and reusable packaging. However, in most applications, plastic containers have largely replaced wooden containers because they are more cost-effective, easier to clean for reuse, and do not risk contaminating foods with slivers. Nevertheless, wood as a material is more renewable and after use represents a lower impact on the environment than plastic [22].

3.2 Paper

Paper-based packaging accounts for more than 30% of the global packaging market and is widely used for the packaging of food products due to its broad range of capabilities at a low cost [23]. The most important aspect of paper packaging is that their inert nature extends the shelf life of packed products. Furthermore, paper and paperboard are easy to recycle and environmentally friendly. This kind of packaging is extensively used for packing a variety of milk and soft drink products, baked goods, and a lot more. The standard paper, on the other hand, cannot be used frequently due to the reduced moisture barrier, which leads to microbial contamination. Furthermore, it lacks mechanical strength, which is required for transportation and storage [24]. There have also been several reports that show safety concerns and toxicity risks. In the year 2004, Ozaki et al. [25] conducted a study involving chemical of concern analysis present in paper packaging. The research stated that abietic acid (AA) and dehydroabietic acid (DHA) of contaminants usually to be reported in paper packaging were found to be toxic and potentially genotoxic at higher concentrations. The food contamination risk by the use of paper-based material in its packaging remains on the table, despite the few reported studies of some sort of paper in a treated form that is widely used for various levels of product packaging including foods [24].

3.3 Polymers

Food packaging materials have evolved over time, from primitive tree barks to modern plastic packages. Polymeric materials are quite diverse and adaptable. They can be soft or stiff, transparent, or opaque, thermosets, or thermoplastics and can be made into films or containers of various shapes and sizes. They are, on average, much less expensive and, without a doubt, much lighter. Countless polymers have been chemically synthesized or extracted from natural sources for a variety of applications. The enormous range of development in polymers necessitated their classification on various bases, the most important of which were origin and stiffness [26].

3.3.1 Synthetic polymers

Nowadays, thanks to their predefined properties, polymers are becoming more popular in food packaging. Their desirable performance of mechanical strength thermal stability and, as good barriers to gases and aromatic compounds, synthetic-based polymers are in high demand as packaging materials. The main reasons for the popularity of such polymers are their low cost and widespread availability. Namely, polyolefins are one of the most common plastics used in food packaging. Polyethylene and polypropylene stand out among the finest polyolefin categories owing to their lightweight, processability, flexibility, strength, reusability, and resistance to moisture and chemicals [27].

3.3.1.1 Polyolefins

Polyolefins (POs) are by far the largest class of synthetic polymers made and used today. PO is a collective term for multiple types of plastics: polyethylene (PE) and polypropylene (PP), the two most frequently used plastics in food packaging industries [6, 28]. Several factors have been principally responsible for the great success that POs have enjoyed: low cost of production, lightweight, and broad range of mechanical and chemical resistance properties. The combination of all of these factors has led to the enormous number of ways in which polyolefins are now being so vast in our lives [29].

3.3.1.1.1 Polyethylene

Polyethylene (PE) is the most basic plastic in terms of structure, and it is made by additional polymerization of ethylene gas in a high-temperature pressure reactor. Depending on the temperature, pressure, and catalyst of polymerization, a variety of low, medium, and high-density resins is produced. The processing conditions determine the properties of the final outcome [30].

The advancement of PE and its derivatives has revolutionized the market, allowing the plastic to better compete with glass bottles. Polyethylene-based materials are currently preferred for milk and juice bottles, grocery, retail, trash bags, as well as bread and frozen food bags. Polyethylene is a heat-sealable material that can be formed into tough films with good moisture and water vapor barrier. Furthermore, when heat resistance is required for packaging, so polypropylene-based materials opt. Nonetheless, when compared with other plastics, they do not provide in particular a high barrier to fats, oils, or gases [23].

Low-density polyethylene (LDPE) and high-density polyethylene (HDPE) are the two of the most commonly used polyethylenes in the food industry. LDPE is nontoxic, stretchable, and shrinkable. It is a good moisture barrier, but it has low oxygen permeability and is ineffective as an odor barrier. It is widely used for bags as well as coating papers or boards because it is less expensive than most films. Because it is stronger, thicker, less flexible, and brittle, HDPE is a better barrier to gases and moisture than LDPE. HDPE packaging is waterproof and tear and puncture-resistant [30].

3.3.1.1.2 Polypropylene

Polypropylene (PP) is a low-cost polymer that is catalytically synthesized from propylene. It has good advantages such as good impact resistance, transparency, excellent mechanical property, high melting point, and low density, making it ideal for a wide range of applications. Also, PP has a respectable degradability among polyolefines. Experiments, however, revealed that when used in room-temperature applications, PP has excellent and desirable physical, mechanical, and thermal properties. The PP material has some drawbacks as well, including low-temperature standing, brittleness, and poor aging resistance [31, 32].

Oriented PP is a clear, glossy film with high tensile and puncture resistance. It has a moderate permeability to gases and odors and a higher barrier to water vapor, so humidity changes are unlikely to affect it. It stretches, but not as much as polyethylene While the properties of PP and PE are similar, there are some differences. Lower density, a higher softening point, and greater rigidity and hardness are among these characteristics [33]. It is used in applications that are similar to those of LDPE [34]. It is most commonly found in the packaging of biscuits, snacks, and dried foods [35].

Although these are used primarily for food applications, there are numerous reasons for their success and rapidly increasing market share in packaging technology. Yet, the majority disadvantage of this kind of polymers produced is their poor biodegradability. Gradually, they contributed significantly to a major source of waste accumulation after being used, thus getting involved in environmental hazards. Hence, the incompatibilities with those synthetic materials for packing small volumes of packages necessitate the search for another viable option [36].

3.3.2 Health and environmental impacts

While food packaging is an important part of the food industry and helps to store foods properly, it can also be a source of concern for food safety. When heated, certain packaging materials, such as certain types of plastic, polythenes, and so on, can release toxins, posing a health risk to consumers. A variety of substances are used in food packaging, including dyes for printing interesting and colorful labels and adhesives to keep packaging closed. To protect consumers effectively, the relevant authority certifies each of these food packaging materials independently, subjecting them to stringent testing standards.

Over the last few decades, the food industry is growing exponentially. On the flip side, the impact of industrial development needs to be analyzed. The materials used in food packaging have a negative impact on the environment. The synthetic polymers, including plastics, and their specific use in the food packaging industry posed a burden on annual solid waste generation. In the current scenario, solid waste management is a noteworthy challenge to a human in concern of the environment, and plastic-based polymers waste is a leading one and remains associated with several ecological issues. The more critical increase in packaging material usage and failure of the natural recycling system, along with the high cost of conventional packaging materials, all prompted to look for alternatives [37]. Natural polymers are the first choice for packaging need to explore and utilize on a large scale. They combine the available packaging materials to be associated with minimal damage, and the latest inventions, including, of course, biodegradability.

3.3.3 Natural polymers

A lot of emphasis is given in recent times to create new, more productive, eco-friendly content and develop packaging material compatible with food products. The use of natural polymers is a key major packaging material for the future. Biopolymers for packaging applications were developed because they were not only biodegradable but also divisible and displayed additional advantageous properties similar to customized applications. They are exceptionally well suited for the use of advanced packaging technologies such as active, intelligent, and modified atmosphere packaging [38]. While the concept of sustainable packaging is also prevailing, recent investigations are, therefore, focused on procuring materials from bioresources, and their utility in the synthesis of natural polymeric choices to taking over and replacing the chemical-based ones. Biopolymers are made from bio-based resources, though the bio-based resource content varies in practice [39]. Among the polymers derived from natural resources, bioplastic is a leading candidate.

3.3.3.1 Bioplastics

Nowadays, bioplastics, whose building components are originated from renewable raw materials, have become more popular. These products deal with a high proportion of potentials for enhanced natural recycling. Biomass for bioplastics production can be extracted directly from plants or produced by microorganisms in fermentative processes. Some currently produced and applied biopolymers based on renewable resources include Poly Lactic acid, cellulose, and starch, which are biopolymers that are directly obtained from argo-wastes [40]. Being materials produced from agricultural feedstocks, such sustainable concept, enlighten the approach of turning waste into a wealth of resources. However, "bio-based" does not necessarily imply "biodegradable" or "compostable" [41, 42]. Bio-based products contain renewable raw materials that can be replenished through natural processes. Polymers that can be degraded by microorganisms in the environment over time are examples of biodegradable products [43]. Biodegradable plastics include compostable bioplastics. As a matter of fact, while all compostable bioplastics are biodegradable, not all biodegradable bioplastics are compostable.

Although these materials are environmentally friendly and easy to recycle, they have a number of disadvantages. Physical protection and mechanical strength for transportation and storage are the essential requirements for packaging. The main issue with these innovative materials if used alone for packaging purposes, biopolymers, or bioplastics shows some limitations in terms of functionality: their poor water barrier properties, brittleness, high vapor permeability, and low heat resistance [44–46]. Thus, biopolymers are strengthened with fillers to enhance their mechanical properties, barrier properties, and heat outstanding [44, 45]. Additionally, material product compatibility is an important factor to consider before commercial application. Such products in food packaging provide less evidence for the large-scale commercial claim.

3.3.3.1.1 Bioplastic reinforcement

As mentioned above, bioplastics alone have some limitations, including low water, heat resistance, and brittleness. Interestingly, some research has been done on the reinforcement of bioplastics through chemical and physical cross-linking treatments to enhance the strength of bioplastics. Quite a lot of potential additives can be used as fillers for bioplastics; these additives are in the micro to nano-sized form. Such fillers can boost the mechanical properties, barrier properties, and heat resistance of bioplastic composites compared with those of virgin bioplastics [44]. Although filler reinforcement can greatly improve the bioplastic performance, the environmental and human health safety concerns posed by these materials during their application should not be forgotten [47, 48].

4. Prospects of biocomposites for food packaging

A polymer composite is a multiphase material in which reinforcing fillers are integrated with a polymer matrix, resulting in synergistic mechanical properties that neither component alone can achieve [49]. Biocomposites are composite materials composed of natural fiber and biopolymers such as polysaccharides. Plant-derived fiber and bio-derived plastics resemble biopolymer or bioplastic; these kinds of composites are sometimes referred to as green composites and apparently exhibit more friendliness. Plus, they include biofibers and matrix polymer systems as the combination of two or more biopolymers may result in the creation of a new biopolymer tailored to specific needs and with brand new or more enhanced features [50].

Biocomposite packaging options can be made from a variety of natural sources, including polysaccharides. In current times, the polysaccharide family has managed to develop as novel origin materials as a replacement for their nonbiodegradable petrochemical-based counterparts. The ability to produce and contribute to the formation of a variety of polysaccharide films by imparting hardness, viscosity, and gel-forming capacity. The nontoxic properties allow for biodegradation and do not produce harmful by-products to the environment. It also has excellent gas permeability properties, which extend the product's shelf life [49].

4.1 Polysaccharides

Biopolymers used as raw materials in the manufacture of biodegradable films should be renewable, abundant, and cost-effective. In some cases, they can be derived from waste. Polysaccharides such as starch, cellulose, and chitosan are among the materials being researched under consideration for biodegradable packaging films and biocomposites. These polysaccharides can procedure films with good barrier properties against gas exchange including oxygen and carbon dioxide. On the other hand, tensile strength and elongation percentage are the important mechanical properties because their desirable values are needed to maintain the integrity of the packed food. The tensile strength values that have been showed by polysaccharide-based films differ, but actually some of them exhibit similar values to those noticed in synthetic polymers. Tensile strengths of films based on high amylose starch or chitosan, for example, are comparable to high-density polyethylene films. The major area of concern is the elongation percentage values, which are significantly lower than those observed in synthetic polymers. Accordingly, researchers are looking into combining polysaccharides with other materials to improve the barrier and mechanical properties of biopolymers that could replace synthetic polymers [51].

4.1.1 Starch

Starch is a polysaccharide that naturally accumulates in plants. Potatoes, corn, and wheat are the primary sources of starch for commercial production. Starch contains two major polymeric constituents: amylose in straight chains and amylopectin with highly branched glucose chains [52]. Starch on its own is brittle and incapable of forming films with the desirable properties. For product packaging, starch-based materials have poor mechanical properties. Unless it is plasticized with other materials, chemically modified, or modified with a combination of such forms of treatment, it lacks high elongation, tensile, and flexural strength. Rewardingly, glycerol is the key player used as a plasticizer [53].

4.1.2 Cellulose

Cellulose and its derivatives are among the most abundant and widely used polymers in the packaging industry. Cellulose is a very important polymer in the food industry. It is the well-known biodegradable polysaccharide from which cellophane film can be made. There are numerous natural sources of cellulose, such as biowastes and agricultural wastes, which explains why it is readily available and at competitively low prices. From 1839 to 2021, all past and present studies have introduced the discovery of cellulose and its numerous functional properties; its use in the preservation of fruits and vegetables and in the preparation of various types of composite films notably [54].

4.1.3 Chitosan

After cellulose, chitosan is the world's second most abundant polysaccharide. Because this biopolymer is primarily derived from waste products in the shellfish industry, commercial supplies are currently plentiful, and as a result, it is reasonably priced. Chitosan is a biocompatible, biodegradable, and nontoxic material that is ideal for packaging films. Chitosan is more versatile than chitin due to its structural properties and the ability to create films with different properties and barriers [55]. This biopolymer has been demonstrated to have excellent film and coating properties, as well as an inherent antimicrobial property. As a result, many chitosan-based films have been fabricated and applied in the food packaging industry. However, its high-water sensitivity causes a loss of barrier properties, limiting its industrial application for packaging. Blending this polysaccharide with other more water-resistant polymers has proven to be a viable solution to the solubility problem. Furthermore, it has been demonstrated that reinforcing this biopolymer with fillers can result in novel composites with improved physical properties, such as water resistance, without sacrificing biodegradability.

5. Conclusions

It is common knowledge that every human activity has an impact on the environment. The massive increase in demand for food and food products contributed to the refinement of current synthetic materials as well as the development of more useful and safe ones. The use of biopolymers and their development are one step toward green packaging. Several biopolymers were developed to achieve this goal adapting new avenue in eco-friendly food packaging. Such a change would undoubtedly reduce the use of packaging, a single-use material, minimizing the generation of waste, hence leading to more ecologically responsible packaging materials and achieving more clean processes. Bioplastics have considerable potential as replacements of fossil-based plastics in many applications; they have been applied in several food packaging industries. Furthermore, bio-based materials aid in the transformation of primary by-products into value-added by-products, particularly when residual biomass is used as a raw material in biocomposites. This is a significant growing opportunity for developing countries where agriculture is one of the main economic activities, with agriculture playing a critical role in the development of strategies toward more equitable markets from the standpoint of sustainable development. Furthermore, biomass-derived compounds have a lower overall environmental impact. Finally, the high versatility of biocomposite materials allows for the development of customized packaging materials for various products, as well as intelligent alternatives with enhanced properties, ensuring the product's quality, wholesomeness, integrity, and safety for a longer period of time. In this regard, there is a global trend toward developing materials with optimized properties in terms of product preservation, resource use, and waste generation throughout the entire cradle-to-grave path. In conclusion, the development of green biocomposites for the packaging industry has the potential to address modern society's need for sustainability.

Acknowledgements

Author LH has to start by thanking Dr. Irene Samy, from reading early drafts to giving her advice on the chapter. Thanks are also due to SESC and NilePreneurs—Nile University for facilitating such opportunity.

Notes/thanks/other declarations

Author LH thanks Medhat Benzoher, for his works guiding her even after he has gone. This helped her feel she is doing it right. He was as important to this chapter getting done as she was.

Author details

Liqaa Hamid and Irene Samy*
Innovation, Entrepreneurship and Competitiveness Centre (IECC), Nile University, Sheikh Zayed City, Egypt

*Address all correspondence to: isamy@nu.edu.eg

References

[1] Cachon R, Girardon P, editors. Gases in Agro-Food Processes. UK: Academic Press; 2019

[2] Sharangi AB, Datta S, editors. Value Addition of Horticultural Crops: Recent Trends and Future Directions. India: Springer; 2015. pp. 1-342

[3] Sinha NK, Sidhu J, Barta J, Wu J, Cano MP, editors. Handbook of Fruits and Fruit Processing. US: John Wiley & Sons; 2012

[4] Risch SJ. Food packaging history and innovations. Journal of Agricultural and Food Chemistry. 2009;**57**(18):8089-8092

[5] Robertson GP. Soil Greenhouse Gas Emissions and their Mitigation. Encyclopedia of Agriculture and Food Systems. San Diego: Elsevier; 2014. pp. 185-196

[6] Goswami TK, Mangaraj S. Advances in polymeric materials for modified atmosphere packaging (MAP). In: Multifunctional and Nanoreinforced Polymers for Food Packaging. UK: Woodhead Publishing; 2011. pp. 163-242

[7] Pongrácz E. The environmental impacts of packaging. Environmentally Conscious Materials and Chemicals Processing. 2007;**3**:237-278

[8] Brody AL, Bugusu B, Han JH, Sand CK, McHugh TH. Innovative food packaging solutions. Journal of Food Science. 2008;**73**(8):107-116

[9] Yam KL, Takhistov P. Sustainable packaging technology to improve food safety. IBM Journal of Research and Development. 2016;**60**(5/6):9-1

[10] Etcheverry M, Barbosa SE. Glass fiber reinforced polypropylene mechanical properties enhancement by adhesion improvement. Materials. 2012;**5**(6):1084-1113

[11] Rodríguez-Rojas A, Ospina AA, Rodríguez-Vélez P, Arana-Florez R. ¿What is the new about food packaging material? A bibliometric review during 1996-2016. Trends in Food Science and Technology. 2019;**85**:252-261

[12] Wyrwa J, Barska A. Packaging as a source of information about food products. Procedia Engineering. 2017;**182**:770-779

[13] Cholewa-Wójcik A, Kawecka A, Ingrao C, Siracusa V. Socio-economic requirements as a fundament of innovation in food packaging. Journal of Entrepreneurship, Management, and Innovation. 2019;**15**(1):231-256

[14] Knorr D, Watzke H. Food processing at a crossroad. Frontiers in Nutrition. 2019;**6**:85

[15] Jones JM. Food processing: Criteria for dietary guidance and public health? Proceedings of the Nutrition Society. 2019;**78**(1):4-18

[16] Hugenholtz J, Hylckama Vlieg J. Improving the flavour of cheese. A volume in Woodhead Publishing Series in Food Science, Technology and Nutrition. In: Monitoring Cheese Ripening: New Developments. US: CRC Press; 2007. pp. 351-369

[17] Guillard V, Gaucel S, Fornaciari C, Angellier-Coussy H, Buche P, Gontard N. The next generation of sustainable food packaging to preserve our environment in a circular economy context. Frontiers in Nutrition. 2018;**5**:121

[18] Kehinde BA, Majid I, Hussain S, Nanda V. Innovations and future trends in product development and packaging technologies. In: Functional and Preservative Properties of Phytochemicals. UK: Academic Press; 2020. pp. 377-409

[19] Díaz-Montes E, Castro-Muñoz R. Edible films and coatings as food-quality preservers: An overview. Food. 2021;**10**(2):249

[20] Aviat F, Gerhards C, Rodriguez-Jerez JJ, Michel V, Bayon IL, Ismail R, et al. Microbial safety of wood in contact with food: A review. Comprehensive Reviews in Food Science and Food Safety. 2016;**15**(3):491-505

[21] Fengel D, Wegener G. Pulping processes. In: Wood Chemistry, Ultrastructure and Reactions. Berlin: Walter de Gruyter; 1989

[22] Fink R, Filip S, Oder M, Jevšnik M. Wood in food industry—Potential applications and its limitations. Microbial Pathogens and Strategies for Combating Them: Science, Technology and Education. 2013;**4**:188-194

[23] Kirwan MJ, editor. Paper and Paperboard Packaging Technology. US: John Wiley & Sons; 2008

[24] Verma MK, Shakya S, Kumar P, Madhavi J, Murugaiyan J, Rao MVR. Trends in packaging material for food products: Historical background, current scenario, and future prospects. Journal of Food Science and Technology. 2021;**58**(11):1-14

[25] Ozaki A, Yamaguchi Y, Fujita T, Kuroda K, Endo G. Chemical analysis and genotoxicological safety assessment of paper and paperboard used for food packaging. Food and Chemical Toxicology. 2004;**42**(8):1323-1337

[26] Ruchir P, Kumar B, Deeba F, et al. Food packaging: Natural and synthetic biopolymers. In: Encyclopedia of Polymer Applications. US: CRC Press; 2018

[27] Jetten J. Quality and safety aspects of reusable plastic food packaging materials: A European study to underpin future legislation. Food Additives and Contaminants. 1999;**16**(1):25-36

[28] Marsh K, Bugusu B. Food packaging—Roles, materials, and environmental issues. Journal of Food Science. 2007;**72**(3):R39-R55

[29] Craver C, Carraher C, editors. Applied Polymer Science: 21st Century. UK: Elsevier; 2000

[30] Cruz RM, Rico BP, Vieira MC. Food packaging and migration. In: Food Quality and Shelf Life. UK: Academic Press; 2019. pp. 281-301

[31] Maddah HA. Polypropylene as a promising plastic: A review. American Journal of Polymer Science. 2016;**6**(1):1-11

[32] Shubhra QT, Alam AKMM, Quaiyyum MA. Mechanical properties of polypropylene composites: A review. Journal of Thermoplastic Composite Materials. 2013;**26**(3):362-391

[33] Cacciari I, Quatrini P, Zirletta G, Mincione E, Vinciguerra V, Lupattelli P, et al. Isotactic polypropylene biodegradation by a microbial community: Physicochemical characterization of metabolites produced. Applied and Environmental Microbiology. 1993;**59**(11):3695-3700

[34] Graham Bowditch T. Penetration of polyvinyl chloride and polypropylene packaging films by *Ephestia cautella* (Lepidoptera: Pyralidae) and *Plodia interpunctella* (Lepidoptera: Pyralidae) larvae, and *Tribolium confusum* (Coleoptera: Tenebrionidae) adults. Journal of Economic Entomology. 1997;**90**(4):1028-1031

[35] Hirsch A. Flexible Food Packaging. New York: Van Nostrand Reinhold; 1991

[36] Müller P, Schmid M. Intelligent packaging in the food sector: A brief overview. Food. 2019;**8**(1):16

[37] Sharma C, Dhiman R, Rokana N, Panwar H. Nanotechnology: An untapped resource for food packaging. Frontiers in Microbiology. 2017;**8**:1735

[38] Wrona M, Nerín C. Analytical approaches for analysis of safety of modern food packaging: A review. Molecules. 2020;**25**(3):752

[39] Vert M, Doi Y, Hellwich KH, Hess M, Hodge P, Kubisa P, et al. Terminology for biorelated polymers and applications (IUPAC Recommendations 2012). Pure and Applied Chemistry. 2012;**84**(2):377-410

[40] European Bioplastics. Applications for Bioplastics. 2019. Available from: https://www.european-bioplastics.org/market/applications-sectors/

[41] Hong LG, Yuhana NY, Zawawi EZE. Review of bioplastics as food packaging materials. AIMS Materials Science. 2021;**8**(2):166-184

[42] Van den Oever M, Molenveld K, van der Zee M, Bos H. Bio-based and biodegradable plastics: Facts and figures: Focus on food packaging in the Netherlands. Report No. 1722. Wageningen Food and Biobased Research; 2017

[43] Niaounakis M. Biopolymers: Applications and Trends. NY: William Andrew; 2015

[44] Mohanty F, Swain SK. Bionanocomposites for food packaging applications. In: Nanotechnology Applications in Food. UK: Academic Press; 2017. pp. 363-379

[45] Talegaonkar S, Sharma H, Pandey S, Mishra PK, Wimmer R. Bionanocomposites: Smart biodegradable packaging material for food preservation. In: Food Packaging. UK: Academic Press; 2017. pp. 79-110

[46] Kim YT, Min B, Kim KW. General characteristics of packaging materials for food system. In: Innovations in Food Packaging. UK: Academic Press; 2014. pp. 13-35

[47] Huang Y, Mei L, Chen X, Wang Q. Recent developments in food packaging based on nanomaterials. Nanomaterials. 2018;**8**(10):830

[48] Rezić I, Haramina T, Rezić T. Metal nanoparticles and carbon nanotubes—Perfect antimicrobial nano-fillers in polymer-based food packaging materials. In: Food Packaging. UK: Academic Press; 2017. pp. 497-532

[49] Wang G, Yu D, Kelkar AD, Zhang L. Electrospun nanofiber: Emerging reinforcing filler in polymer matrix composite materials. Progress in Polymer Science. 2017;**75**:73-107

[50] Tajeddin B. Cellulose-based polymers for packaging applications. In: Kumar Thakur V, editor. Lignocellulosic Polymer Composites. Beverly: Scrivener publishing; 2014. pp. 477-498

[51] Cazón P, Velazquez G, Ramírez JA, Vázquez M. Polysaccharide-based films and coatings for food packaging: A review. Food Hydrocolloids. 2017;**68**:136-148

[52] Coma V. Polysaccharide-based biomaterials with antimicrobial and antioxidant properties. Polímeros. 2013;**23**:287-297

[53] Bertuzzi MA, Vidaurre EC, Armada M, Gottifredi JC. Water vapor permeability of edible starch based films. Journal of Food Engineering. 2007;**80**(3):972-978

[54] Liu Y, Ahmed S, Sameen DE, Wang Y, Lu R, Dai J, et al. A review of

cellulose and its derivatives in biopolymer-based for food packaging application. Trends in Food Science and Technology. 2021;**112**:532-546

[55] Fernández-Saiz P, Sánchez G, Soler C, Lagaron JM, Ocio MJ. Chitosan films for the microbiological preservation of refrigerated sole and hake fillets. Food Control. 2013;**34**(1): 61-68

Chapter 9

FRP for Marine Application

Bikash Chandra Chakraborty

Abstract

Fiber Reinforced Plastics (FRPs) are widely used in marine sector owing to their high specific strength and resistance to marine corrosion. For naval application, additional advantages are transparency to radar wave and better vibration damping than metals. The use of various FRPs in off-shore structures and marine vessels needs analysis of desired properties considering the types of matrices and fiber. The common consideration is effect of sea water on the properties of the FRP. This chapter gives a brief on use of different FRPs in various areas such as off-shore pillars, Reinforced Cement Concrete (RCC) enclosers, primary and secondary marine components. A brief discussion is included here on diffusion models and estimation of durability by a time-temperature superposition principle applied to water ingress and corresponding change in mechanical strength of FRPs with examples. The effect of microbial activity on the damage of FRP is not very much reported in literature. It is known that sulfate-reducing bacteria (SRB) are the most damaging microbes for FRP. In conclusion, it is highlighted that vinyl-ester-based FRPs using glass and carbon fibers are best for marine application. To determine the realistic service life in marine environment, Vinyl Ester- FRP (VE-FRP) are to be simultaneously studied for damage due to sea water and the microbes such SRB.

Keywords: GFRP, CFRP, epoxy, unsaturated polyester, vinyl ester, marine environment, sea water diffusion, time-temperature superposition, life time, microbial activity, sulfate-reducing bacteria

1. Introduction

Use of Fiber Reinforced Plastics (FRPs) is rapidly expanding in all fields such as medical equipment, engineering plants, packaging, transportation, aviation, space technology, building construction, heavy vehicles, and defense forces. Application of FRPs in marine construction industry is also not new and ever increasing with rapid advancements in exotic fibers, nanoparticles, and special polymers. For engineering application, principal requirements are inherent strength and a defined temperature limit up to which the strength is sustained to the desired level. The secondary requirements are high toughness, resistance to cyclic fatigue, low creep, low relaxation, environmental stability, and ease of joining and maintainability. The third most important factor is investment cost and processing cost. So far as mechanical strength is concerned, the Elastic Modulus in all modes and ultimate strengths are important. However, too stiff composites lack toughness, which often cause premature brittle failure. It has to be a tread-off between ultimate property and elastic modulus for restricting strain on the one hand and sustain low/high cycle fatigue on the other hand. The toughness imparted by flexible long-chain resin matrix results in high creep and relaxation, which are undesirable for engineering structures but improve the fatigue life. Inclusion of rubbery moieties in a stiff matrix may result in phase separation and stress concentration at the interface and may cause premature failure. Toughening by nanoparticles, such as functionalized carbon nanotubes and reduced graphene oxide and derivatives, are being actively researched at present with apparently encouraging results. Detailed study of creep and stress relaxation of CNT-polymer or graphene-polymer composites is not done yet in a comprehensive manner, but with a general understanding, it is expected to be even better than the pristine polymer.

Thermal properties are more extensive, since the thermal agitation of polymers undergoes very drastic rise beyond a characteristic temperature called glass transition, where the stiff polymer transforms into a rubbery soft material. For polymers with partial crystallinity, flow takes place at further enhancement of temperature, and finally, a polymer starts to decompose at even higher temperature. A design of structural element then has to depend on the limit of temperature at which the modulus starts decreasing. Ideally it should be glass transition temperature. However, in practice, dynamic mechanical analysis shows that the modulus decreases even about 10–15°C below the glass transition. The extreme hazards of heat for an organic polymer (and FRP) are the fire propagation and evolution of toxic gases. The fire-retardant additives both as physical addition and chemical modification of resins are widely used and are also currently being researched in the light of possible benefits of nanoparticle reinforcements.

Marine application both for static off-shore structures and sea-going vessels needs robust and durable FRP composites, which can compete well with metals in terms of specific strength, durability, and cost-effectiveness. The replacement of a metal requires some special properties in FRPs apart from strength and degradation. One of the most difficult solutions is joining the Thermoset FRP elements since the joint should be almost similar in mechanical strength and toughness. Identical thermoset as the FRP element is best preferred, with a short fiber dough molding system that must be cured at ambient, yet provide acceptable joint strength. There can be special drilling technique for joining through riveting using the dough as rivets. Thermoplastics can be "welded" by melt joining as metals, most suitable for particulate reinforced composites and short fiber composites. The second and very important property of an FRP to qualify marine standard is effect of sea water aging considering all the chemical and biological adversaries of the sea. This single factor mostly decides the design and service life of a marine-grade FRP structure.

Although marine corrosion of FRP is not so severe as for steel, the FRP structures and underwater hulls need to be protected from bio-fouling. With the advancement of anti-fouling coatings, it is possible to protect a hull for minimum 3 years without any maintenance painting. Modern low surface energy foul release coatings based on silicones and fluoro-silicones are environment-friendly as they do not release toxins in the sea. These are non-depleting coatings and hence can have higher service life. However, These types of coatings are more effective for high-speed boats.

2. FRP components in marine vessels

The different elements of a ship can be defined as primary, such as superstructure, hull, SONAR Dome, bulkhead, decks, propeller shafts, masts, doors, hatches, machinery foundations, support frames, etc. Secondary items are rudder, pipes, valves, ladder, stanchions, guard rails, etc.

In naval vessels, three important advantages of using FRP composite are (1) ability to damp vibration, thereby reducing the radiated noise in the sea. In addition, FRPs are acoustically transparent, hence reduce the acoustic reflection (2)

FRPs without carbon or conducting material inclusions are radar transparent. These two features enhance the stealthy character of a battle ship and submarine and importantly (3) most common reason is no corrosion of FRP in sea water and saline atmosphere.

The most used application areas for FRP in ships are superstructure and bulkhead, where thick FRP panels are used with flap joint overlapping at the corner to flush the sides. Riveting with composite rivets can be done along with interface adhesion using a hand layup of fabric with resin so that the joint is sufficiently strong.

Vibration and fatigue are other important aspects. Machinery and propeller movement cause vibration of the hull and hull-mounted SONAR dome, which adversely affect SONAR performance and also results in fatigue. Normally, the fundamental frequency of machinery and propeller is up to 34 Hz, and prominent modes are up to about 200 Hz. Also, slow cycle fatigue results from sea waves, which is approximately 0.8 Hz. It is well known that slow cycle fatigue is quite important to decide the service life for steel hulls and is expected to have similar effect on FRP hull. Till now, there is no such detailed study on fatigue at various frequency envelops for marine FRPs, which are actually exposed in sea water with vibrations.

2.1 FRP hull

Hull construction using FRP is very common for speed boats, small to medium size (8–80 m) patrolling boats, research ships for acoustic and underwater mapping studies, coastal ships, corvettes, etc. [1–4]. A very comprehensive list of literature is given by Galanis [3] in his M.S. thesis. One very interesting naval ship is mine countermeasure vessel (MCMV) which is about 60 m long [1]. MCMV uses passive magnetic sensors to detect underwater mines, which use a magnetic sensor to trigger the mine. Therefore, the ship as such should not have any magnetic signature. Conventional hull material of MCMV is nonmagnetic steel. However, FRP is preferred because of lightweight and corrosion-free nature in addition to nonmagnetic character. Miller [4] reported one such MCMV of Royal Navy (U.K.) made of GFRP way back in 1973 and a bigger one (60 m long) in 1980s. FRP is being used since World War II by the United States in noncritical areas and small boats [2]. Till now, large commercial or Naval ships such as Frigates, etc., are not made with FRP. The FRP hull of ships of 60–80-m size is a sandwich construction with thick FRP skin and a foam core. Thickness of the FRP skin on both sides of the foam core can be 8–10 mm each, and the foam core can be 50–80 mm thick. Previously PVC foam core was very common. However, with the large variety of polyurethane foam available today, even fire-retardant type including polyisocyanurate-modified polyurethane [5], the scope and ease of foam filling in between two hull panels have facilitated production system, and also large seamless foam core is easily made by foam spray machine.

Generally, unsaturated polyester made with isophthalic acid and neopentyl glycol and epoxy-based vinyl ester resins is widely used for marine boats and ship hull. Both of these resins are cross-linked by styrene monomer to form a thermosetting polymer.

FRP hull of boats of maximum 8 meters contains six types of fiber and fabric layers arranged in a sequence and requires minimum 25 layers of reinforcing mats and fabrics. The different fiber-based layers are random chopped strand mat, 300–450 GSM, woven roving mat 400–600 GSM, fabric of different thicknesses (approximately 0.25 mm), core mat of 1 mm and 3 mm thickness, and a 1 mm skin layer of the resin with particulate fillers (titanium dioxide, aerosil, barytes, etc.). Aerosil (fumed silica) is used with the resin to make it more impermeable to water. However, aerosil makes the resin somewhat thixotropic. Due to large size, and to make seamless hull, wooden mold is first made, and hand lay-up technique is used for fabrication. In an elaborate arrangement, vacuum bagging or vacuum-assisted resin infusion can be used for at least small boats. Vacuum-assisted molding can make composite with about 70% fiber and 30% resin, which is obviously advantageous for strength.

2.2 SONAR dome

A special application of naval ships, submarines and fishing vessels is SONAR dome, which houses the arrays of acoustic transmitting and receiving transducers for detection of underwater objects. Conventionally titanium is used to make sonar domes due to its fair acoustic transparency, high strength-to-weight ratio, and good resistance to sea water corrosion and bio-fouling. However, acoustic impedance of titanium is not as close to that of sea water compared with glass-fiber based FRPs. Therefore, titanium domes are less efficient in underwater acoustic transmission and have underwater acoustic reflectivity more than FRP. For naval vessels such as submarines and battleships, FRP SONAR domes are being used in some countries, for example, the United Kingdom France, Sweden, Australia, Holland [1–3]. Such SONAR domes are very critical with respect to high drag force, compressive stresses at high depth in sea for submarines, and requirement of high acoustic transmission characteristics. The thickness of the dome is decided by the strength and modulus of the FRP, but higher thickness results in loss in acoustic transmission power. Therefore, the design and fabrication of an FRP dome are very critical and are done using Finite Element Method (FEM) so that the dimensional features, strain levels at different sections, and maximum stress can be somewhat accurately determined for both static and dynamic conditions. A prediction of acoustic transmission can also be done using general acoustic attenuation theories. Fabrication method can be very important, so that the dome would have nearly same theoretical strength and dimensional accuracy with acceptable tolerances. Among many possibilities, resin film infusion technique or vacuum-assisted resin infusion can be adopted to make the domes with precise dimensions, strength, and flawless integrity. The thickness of such domes can vary from 20 mm to 80 mm depending on the size. The vacuum process has two advantages, (1) the high fiber content resulting in high strength and (2) nearly zero air gap/flaw in the composite. The air gap is undesirable in sonar dome since any such air bubble would increase acoustic reflection, thus reducing the acoustic transmission across the dome thickness. The aspect of sea water diffusion and corresponding loss of strength, lowering of glass transition, and deterioration of acoustic transparency are main consideration of its long usability and depend on both material and fabrication process. Commonly used fabric is E-glass and S-glass while the resin can be a hybrid of vinyl ester and epoxy resin. For the purpose of enhancement of strength, carbon fibers are preferred over glass fiber, since a carbon fabric-epoxy FRP would have Young's modulus of nearly 70 GPa compared with GFRP of about 30 GPa.

Recently aligned carbon nanotube containing GFRP domes are being considered for mid-frequency acoustic application to reduce the thickness of the dome and to impart better structural vibration damping.

2.3 Secondary marine components

Of the superstructural components, carbon fiber-epoxy combination is best, provided there is no necessity of radar stealth features. However, carbon fiber-epoxy composites are used in cabinets and covers of power electronics in ships and submarines for EMI shielding purpose.

Composite pipe can be made using a combination of prepreg lay-up on a mandrel followed by filament winding technique. This fabrication method can give sufficient Hoop Stress. A best possible fiber alignment in subsequent layers on the mandrel is determined by a stress-strain analysis by FEM method. Resin pick-up by the fiber strands in automatic winding method is minimized by two doctor's blades fixed on the fiber running line as one of the guide systems for the strands before winding onto the mandrel. Autoclave curing at high pressure and temperature can be adopted for such pipes.

Composite valves are made by dough molding compounds because of intricate dimensional requirement to make them leak-proof. Pipes and valves are special among all items because the fluid pressure (Pascal's pressure) in most commercial ships is designed for 12 bar, and for Naval standard, it should be 20 bar with continuous use and should withstand maximum 30 bar pressure for 24 h. This stringent requirement makes the fabrication method very critical, for example, the surface of the pipe must not "sweat" at high hydrostatic pressure and circularity and movement of the ball in a Ball valve must be very precise to avoid leakage of liquid, besides resistance to the "sweating." The processing and fabrication with dough molding compounds are best done by application of high pressure of 1–3 MPa to eliminate excess resin and to ensure compactness with precise dimensional tolerance and without layer gaps or air entrapment. Vacuum application is not beneficial since the dough, containing 20% short fiber, would have very poor flow property. Instead, kneader mixing can produce dough without air entrapped in the green dough. The molds are made with die steel for high-pressure molding.

The elements that are used on board such as ladder, stanchion, and guard rails are critical due to shape and require high-impact energy to resist crack or breakage on impact. Hence, a method of flexibilizing or nanoparticle reinforcement must be attempted to improve impact energy of common reins. As an example, a common epoxy thermoset Glass FRP has an impact energy of 750–850 J/m (Charpy impact), while a modified epoxy-Glass FRP would have 1300–1500 J/m, which may qualify the impact requirement. The strength must not be compromised too much. A maximum 10% reduction for the FRP could be accepted by a designer to prefer a flexibilized resin matrix. For such small and shaped components, hand lay-up of fabric and resin or prepreg lay-up in metallic mold can be adopted. High compression would be beneficial to eliminate any flaw, air gap, and better compactness. In these on-board components, carbon fabric prepregs cannot be used in naval vessels since carbon-based composites increase radar signature.

3. Off-shore structures

Off-shore marine structures such as oil rigs and columns of bridges, underwater pipe line supports, etc., are conventionally made using reinforced cement concrete having steel rods as reinforcements inside the concrete. In some other cases, steel pipes, pillars, and column supports are used.

The underwater static steel structures are protected from corrosion and fouling by electrochemical protection system and paints. A new method of protecting the steel structure is to provide a wrap of composite as an outer lining, which is far more durable than painting and more maintenance-free. Steel pipes are used as a mandrel for a filament winding technique to provide a composite lining. Steel structures require underwater welding, etc., for repair and maintenance, which is very complicated and costly. The FRP lining provides a very convenient solution to reduce such maintenance cost and frequency of repair.

The concrete with diffused sea water generates more alkali, and the pH of this alkaline seawater increases from normal range of 8–8.3 to about 12–13 and accelerates the corrosion of the steel reinforcement in the RCC structure. Therefore, FRP reinforcements are modern way of construction for higher durability and lesser maintenance. However, in a higher alkaline sea water environment, the FRP degradation is expected to be faster than in normal sea water.

For off-shore and maritime civil engineering structures, carbon fiber composites (CFRP) are preferred over glass fiber (GFRP) because of higher mechanical strength of CFRP. In addition, sea water uptake and degradation of GFRP in sea water are higher those in than CFRP. There are very few applications of GFRP in marine structures despite the fact that GFRP is cheaper compared with CFRP.

The durability is also dependent on the resin type and its interface bonding with the fiber. Generally, thermosetting polymers such as epoxy, polyurethane, phenolic resin, vinyl ester, and unsaturated polyester resins are used for composites. These resins and corresponding composites are to be evaluated for long period of sea water exposure in an RCC construction for durability. As accelerated studies might give some extrapolated figures of service life, but such studies cannot determine the effect of microorganisms on degradation of a composite. A very common example is sulfate-reducing bacteria (SRB) in the sea water. These organisms use sulfates dissolved in the sea (for example, $MgSO_4$) for metabolism and produce hydrogen sulfide, which is highly corrosive to metals and may also increase the degradation of composites after settling onto the surface. The effect of such organisms is much more important for static structures rather than moving objects. Fouling by other micro and macro-organisms and subsequently the effect on the composite is another aspect of static structures. The protection from bio-fouling by application of anti-fouling coating is another subject of study. However, this type of coatings work either on toxin release mechanism or by providing a low surface energy coating. In case of toxin release coating, the toxin release depends on the hydrolysis and dissolution of the toxin in water, which is more effective in moving condition than stagnant water. Because of toxin depletion, the coating requires renewal after a certain time, mostly 3 years. In case of low surface energy coating, the effectiveness is far less for static structures, as this type of coating is quite good for moving objects, that too at certain minimum speed. However, the advantage is that the settlement of bio-fouling species on these low surface energy coating is very weak and can be removed by a soft cleaning mop. The second most important aspect of static structures is stress. Most supports and beams are under stress, small or large. The pre-stressed composite structure may have lesser service life compared with no-stress elements. The third consideration is fatigue. A bridge column, pipeline carrying liquids under the sea are subjected to vibrations. Hence, the composite elements are to be evaluated by fatigue for a predetermined frequency and number of cycles. This should be done in a repeated experiment and at a regular immersion period. The effect of pre-stress and vibration parameters may reveal some results, which could be different from normal static experiments.

4. Quality of a marine-grade resin and FRP

A general understanding of large-scale application of FRP is that there can be three main alternatives for a techno-commercially viable thermoset selection, e.g., (1) unsaturated polyester resin (USP) cross-linked with styrene, (2) vinyl ester resin (VE) cross-linked with styrene, and (3) epoxy resin cross-linked with amine. Whereas, there can be two common fibers such as glass and carbon.

Service life of a structural element for marine vessels has to be minimum 25 years for reducing the investment for replacement and should be maintenance-free for at least 8 years to reduce the cost of refits in drydocks. For off-shore structures, where FRPs are used to make barrier for underwater cement concrete structures, the maintainability is even more difficult, requiring high service life without maintenance activity. Apart from the general physical and mechanical properties, an FRP for marine application must have additional characteristics of low moisture/sea water ingress, minimum hydrolysis, good bond strength between fiber and polymer, minimum physical damage of fiber and polymer due to water ingress and retention of mechanical properties even after prolonged sea water immersion. However, all the properties are primarily dependent on the matrix polymer and fiber and their interaction, secondary parameter being processing technique and fabrication methods to make flawless FRP components with fairly accurate dimensions, such as for a marine ball valve or pipe joint. Processing assumes larger importance since partially cured samples are prone to poorer physico-mechanical properties and higher degradation in water.

A significant improvement in properties of marine composites can be achieved by prepreg method and resin transfer molding (RTM) assisted by vacuum. Good compaction and high fiber volume fraction can be achieved by these processes. The process of prepreg molding is feasible where the resin-hardener reaction does not take place at ambient or storage temperature, and the curing is done at a fixed higher temperature. There are high-temperature reacting systems such as epoxy resin 4,4′methylene dianiline tetraglycidyl ether (TGDDM), to be cured with hardener such as 4,4′-diaminodiphenyl sulfone (DDS), modified polyamines, etc., which are used for making prepregs. The shelf life of such prepregs at storage temperature of −20°C is about 10–12 months, but few weeks at 20°C. The prepregs are cured in compression at above 100°C. However, prepreg system may not be possible for vinyl ester or polyester resins. RTM process requires low viscosity resin and hardener to facilitate good flow in the fabric stacked in the mold for proper wetting at all corners and contours. Trujillo et al. [6] reported the properties of RTM processed composites based on three common resins, i.e., epoxy, vinyl ester, and unsaturated polyester with glass and carbon fabrics. The flexural modulus of the glass composites was about 40GPa and about 110–120 GPa for carbon fabric composites, while the flexural strengths were seen to be in the range of 600–800 for glass composites and 1300–1400 MPa for carbon fabric composites.

Sea water absorption causes changes in the matrix by both plasticization and hydrolysis. Initial effect of water ingress is a plasticizing effect and swelling of the polymer matrix. The results are lowering of glass transition temperature due to plasticization and a possibility of debonding of the polymer-fiber interface due to swelling of the polymer. The initial effect of water ingress also causes hydrolysis of the fiber sizing and generates alkali ($Na+$ and $K+$) and the Fiber-polymer interface weakens. All these events result in reduction of ultimate strength and elastic modulus of an FRP.

On prolonged exposure, several chemical reactions may take place, such as hydrolysis of the polymer resulting in small molecules such as glycol, chain breaking, and release of low-molecular-weight polymer (especially polyester), release of the constituents of the resin (typically maleic/fumaric acid), release of styrene (cross-linker for polyester and vinyl ester), and extraction of these species from the FRP to the sea water. Prolonged water immersion of FRP may also cause mechanical damage to the fiber and polymer both, which may not be observed in short period, even in few months of exposure. SEM analysis of all FRPs irrespective of the fiber showed detachment of matrix from the fiber, which is the main reason for such drastic decrease in strength of the composite laminates as reported in literature.

The polymer plays the most important role in the hydrolytic durability of an FRP. As a special case of glass fiber reinforcement, the coated material used as coupling agent chemically degrades and causes weak interface of fiber-polymer.

Therefore, a polymer-fiber combination is ultimately the consideration for optimization of hydrolytic properties.

The main reason why GFRP is not used in maritime civil construction applications is because sea water environment degrades the long-term mechanical properties of GFRP composites and interlaminar shear strength (ILSS). The glass fiber-polymer interface is strengthened by a coupling agent coated on the glass fibers and the process is called "Sizing." The sizing formulations are very complicated, may contain many different chemicals, and are proprietary to the manufacturers [7]. Most common are γ-amino propyl tri ethoxy silane (APTES), γ-glycidoxy propyl trimethoxy silane (GPTMS), γ-methacryloxy propyl trimethoxy silane (MPTMS), and vinyl tri ethoxy silane (VTES) having Si-OH groups on the fiber surface for improving the interface adhesion with the resin. The sea water diffused to the interface of fiber and polymer very quickly degrades the glass to produce alkaline oxides unless protected by sizing. Even then, prolonged immersion of the GRP with fiber with appropriate sizing may cause leaching of alkali oxides (sodium and potassium) from the surface of the fiber and degrade the composite mechanical property [8].

4.1 Epoxy thermoset-based FRP for marine application

Epoxy resin is a versatile thermoset, widely used in many marine structures for many years. It has good mechanical properties, is highly polar and compatible to most fibers including metals, glass, carbon, Kevlar, and polybenzimidazole. Epoxy nanocomposites are gaining importance due to lightweight and high performance in some functional properties when used with carbon nanotubes, nanofibers, graphene, and also natural nanofibers. The conventional epoxy resin thermosets are somewhat brittle and, in many occasions, modifications are done either by physical mixing or chemical reaction onto the epoxy oligomer or use of high-molecular-weight epoxy and/or the amine curing agent to make optimum tough thermoset. However, flexibilization means increase in free volume in the polymer and subsequent increase in moisture absorption. As such the degradation of conventional epoxy thermoset and composites is very widely studied by many researchers since last 45 years, for example, by Augl and Berger [9] in 1976 on carbon fiber-epoxy composites, McKague et al. [10], DeIasi and Whiteside [11], and Whitney and Browning [12] studied moisture diffusion in epoxy matrix and composite, during 1976–1978, to name a few. Similarly, Loos and Springer [13], Bohlmann and Derby [14], Shirrell [15] studied moisture diffusion and its effect on graphite epoxy composites way back during 1976–1979. Glass fiber-epoxy composites are most widely evaluated for effect of moisture or water or sea water absorption from those periods and are still being the subject of study. A few are listed here as references [16–30].

The glass transition temperature of cured epoxy matrix and composites is reduced from 120°C to as low as 66°C on a 2-month sea water exposure, but was observed to be almost constant around 85–88°C from 4 months onward till the end of the study period (12 months), as reported by Chakraverty et al. [31]. The authors explained this anomaly by probable osmotic effect of the bulky molecules of dissolved salts in sea water, which might have initially facilitated the creation of more free volume in the cross-linked epoxy matrix, but on prolonged exposure, deposition of these salts could have reduced the water ingress. Murthy et al. [32] have shown that the water uptake by epoxy-glass composite is more (about 0.9%) compared with 0.7% by epoxy-carbon composites after 12 months and remained unchanged. However, their study was limited to 16 months. The ILSS was reduced by 38% for epoxy-carbon and by 31% for 450 days at ambient temperature. SEM analysis revealed that the moisture penetration along the fiber/matrix interfaces caused interfacial debonding and consequently degradation of the interface. Espinel et al. [28] also showed that for an epoxy-glass composite, the saturation level of sea water was 0.4% at 25°C attained after 30 days. The tensile and flexural strength reduced by about 24% and 35% respectively after 90 days sea water immersion at 25°C, but observed that the strength did not decrease much after saturation of sew water. Contrary to these results, Murad et al. [25] showed that the sea water intake in epoxy-glass unidirectional composite was 2.5% after 12 months, but the strength and elastic modulus had no noticeable change compared with fresh sample. However, the fiber volume % was only 52. Wood and Bradley [20] also reported about 2.2% sea water uptake for 5 months at ambient temperature for an epoxy-glass-graphite hybrid composite, each layer fabricated by a similar process as filament winding, and hence the layers were unidirectional. The glass and carbon were in transverse direction to each other. However, the resin used had a 5% flexibilizer (rubber) and fiber volume % was 60. Komorek et al. [29] used fabrics of glass and carbon in epoxy resin. The bending strength was found to be 8% less for the samples immersed for 36 days at 15°C in sea water.

A unique study on fatigue and sea water aging of epoxy-glass and epoxy-Kevlar composite was done by Menali et al. [33]. The authors studied the effect of sea water (artificial) immersion (40 days) after fatigue for 100–50,000 cycles for these composites. There was about 19% reduction in tensile strength for the Glass-epoxy composite samples and about 15% for Kevlar-epoxy samples which had undergone 50,000 cycles of straining and aged in sea water for 40 days. The stiffness of the composite laminates was also degraded by almost similar extent. This result, when compared with that of Komorek [29], clearly shows the additional degradation under cyclic loading.

There is another interesting review report by Li et al. [34] on effect of alkaline sea water (pH at 12–13) for pre-stressed FRP laminate and FRP tendons. The alkaline sea water simulates the property of the sea water sea sand concrete (SWSSC), which is now very much used in civil construction of marine static structures such as off-shore platforms. The authors compiled several results by some researchers. It is seen from the review article that the alkaline SWSSC at pH of 12–13 has a higher degrading effect under such condition.

A comprehensive study was done on the effect of sea water immersion at various temperatures for an epoxy thermoset plaque and its E-glass fabric composite having 55% fiber by volume. The report is not for publication. The composite samples were made by vacuum bagging process followed by compression molding at 120°C. The curing of plaque and composite was done after thorough degassing of the resin-hardener mix. It was observed that after 360 days of immersion, the flexural strength reduced from about 90 MPa to about 65 MPa, and the dynamic flexural modulus was reduced from about 3.20 GPa to about 2.5 GPa at 30°C in natural sea water. The E-glass composites of the same resin were seen to deteriorate in flexural strength and modulus. The strength reduced from 250 MPa to about 180 MPa, and the dynamic modulus reduced from 8 GPa to about 5.5 GPa. The results clearly show the effect of debonding of the fiber from the epoxy matrix interface thereby drastically reducing the loadbearing capability. The water had a plasticizer effect too, as the glass transition temperature changed from 60 to 62°C to about 52–54°C in 12 months, and the SEM micrograph showed separation of the fiber from the matrix at the interface very clearly. However, the effect of the microbes on degradation could not be quantified separately.

The studies done so far indicate a common observation and conclusion that the degradation of epoxy-based composites is significantly high in terms of delamination, loss of mechanical properties and glass transition on exposure in sea water even for a year. The initial moisture ingress has a plasticizing and swelling effect, due to which the glass transition temperature reduces with a drop in mechanical properties. In prolonged exposure, the water molecules chemically react with the resin (hydrolysis) producing small chemical substances, which tend to diffuse out of the resin, causing blisters. Also, various salt components of the sea water may affect the moisture absorption rate compromising some properties of FRP in sea water. It is also known that the effect of sea water on glass fiber reinforced composites differs according to the type of matrix and fiber. The mode of failure of glass/epoxy composite is altered from a brittle matrix and ductile fiber to ductile matrix and brittle fiber. However, in some opinion, the strength stabilizes after the absorbed moisture attains saturation.

In construction of FRP elements of ships, the items that are not in continuous immersed condition such as superstructures, ladders, stanchions, guard rails, etc., are better designed with toughened epoxy resin and carbon/glass fabric composites since the degradation is limited in the atmosphere and the composites can have sufficient strength, reasonable glass transition temperature even after the toughening process of the resin. For naval ships of stealth features, carbon fiber and nanocarbons cannot be used as the radar reflection will be increased. For elements to be used underwater, epoxy resin is not that superior to the vinyl ester class of resins.

4.2 Vinyl-ester-based FRP composites

Vinyl ester resins are most commonly used for marine composites for two main reasons, the mechanical strength retention on prolonged exposure in sea water and the strength is comparable to epoxy composites and higher than polyester-based composites. The resin has inherent resistance to water diffusion and consequently lesser effect on its glass transition and strength. For large ship structures, vinyl ester resin is a better thermoset due to suitability for processing large items such as hulls, using vacuum infusion due to its low viscosity, apart from its durability in marine environment.

Conventional vinyl esters are having aromatic backbone of epoxy base and the double bond of the unsaturated ester is cured by styrene, exactly the same process as a polyester resin. The presence of the higher content of stiff aromatic epoxy backbone provides the higher mechanical strength compared with phthalic-acid-based polyesters. The higher aromatic content also restricts the diffusion of fluids. Unlike epoxy matrix cured by amines, the vinyl ester matrix is cured by hydrophobic monomer styrene, and hence, the water ingress is lesser than epoxy resin.

VE-CFRP and VE-GRPF have different strength ratios depending on the mode of force application. Wonderly et al. [35] compared these two types of composites in terms of tensile strength and found that CFRP was about 850–950 MPa and was 1.6–1.75 times higher than GFRP, but the open hole tensile strength was comparable at about 250–265 MPa, and compression strength of GFRP was about 330–360 MPa for CFRP and was about 17% lower than GFRP. Transverse tensile strength of CFRP was also about 75% of GFRP. One interesting study was done by the authors on ballistic impact test, which is important for military application. At a comparable areal density, the specific energy (J/kg/m^2) required to penetrate the panels for CFRP was higher by about 25% compared with GFRP for identical muzzle velocity.

In general, the glass transition temperature of a vinyl ester matrix is about 115–120°C. The flexural modulus and strength of a vinyl ester plaque are about 3.0–3.5 GPa and 80–120 MPa respectively, almost same as epoxy plaque. GFRP of vinyl ester has flexural modulus of about 10–12 GPa and flexural strength of about 270–300 MPa, ILSS of about 30–35 MPa, depending on the fiber type and fiber volume fraction in the composite. The CFRP of vinyl ester has much higher strength and modulus, about 3–3.5 times higher than GFRP at the identical volume fraction of fiber.

Water diffusion studies show on an average 0.6% water intake at equilibrium for GFRP and about 0.4% for CFRP, which are almost half of corresponding figures for epoxy CFRP. Murthy et al. [32] showed that the sea water saturation levels in both GFRP and CFRP of vinyl ester are about 0.7% and approximately 0.4% respectively, and there was no reduction in the total weight of the samples even after 450 days of immersion. The interlaminar shear strength was reduced by about 35% after 365 days for both CFRP and GFRP of vinyl ester. Similar extent of degradation was observed for flexural strength. The reduction of tensile strength was about 30% for the same period of immersion. However, it is observed that the mechanical properties and the water uptake almost became steady after 365 days. The authors showed that after immersion in artificial sea water for 450 days, the strength reduced by about 35% for both the composites. Similarly, the ILSS also reduced by about same extent. CFRP is marginally better in ILSS on aging in sea water. The maximum water intake for CFRP was about 0.4% compared with about 0.48% for GFRP.

A study by Mungamurugu et al. [36] showed about 1% water absorption at 20°C for vinyl ester GFRP composite for glass fiber volume of 58% compared with about 0.75% for the plaque after 450 day and the reduction in flexural strength (original 250 MPa) by about 25% for the composite after 300 days.

However, most experiments reported in literature are done with artificial sea water, and the effect of the microorganisms and of the evolved materials due to the metabolism of the microbes present in sea water was not possible to observe. Therefore, the drastic decrease in mechanical strength for thermosets resin plaques is due to reaction with water and hence loss of molecular integrity of the cross-linked matrix.

4.3 Unsaturated-polyester-based FRP composites

Unsaturated polyesters (USPs) are also widely used in marine construction since it is very cost-effective, easy to process by vacuum-assisted resin transfer process due to low viscosity. Easy to cure, and intricate shapes can be made with a large variety of USP. The oligomer resin is cured conventionally by styrene in presence of catalysts such as methyl ethyl ketone peroxide (MEKP) and in some cases added accelerators such as a cobalt salt or those based on a tertiary amine. The cross-link density and corresponding mechanical properties are controlled by styrene content and the unsaturation in the oligomer. There are new USPs developed where styrene is replaced by acrylic monomers such as tri- or tetraethylene glycol dimethacrylate (TEGDM) [37], which are comparatively less toxic than styrene. The general-purpose and marine-grade USPs are synthesized with different glycols such as isomers of pentyl glycol and isomers of phthalic acid with small amount of an unsaturated acid such as fumaric acid or malic acid. The average flexural strength of USP-based GFRP is about 250 MPa with 58–60% glass fiber by volume.

A study on long-term natural sea water immersion of USP was done by Norwood [38]. The USPs of orthophthalic-acid-based marine resin and isophthalic acid-neopentyl glycol (IST-NGP)-based marine-grade resin were used with about 2.25:1 ratio of CSM: resin by weight in the form of chopped strand mat (CSM) and 1:1 ratio by weight for woven roving (WR): resin. The surface tissue coating of the same resin with 5% filler content and a gel coat was used to reduce water permeation. The study revealed that the IST-NGP-glass composite showed best water resistance in terms of appearance of blisters. The best performance was of high HDT (heat deflection temperature) IST-NGP where the blister formation was seen only after 200 weeks, while orthophthalic-acid-based conventional marine resin (medium HDT) showed blisters in about 52 weeks as the best performance. The conclusion in that study was significant for subsequent research on marine-grade FRPs. It was suggested to use a tissue layer of about 5% (by weight) CSM in the IST-NGP resin (high HDT grade marine resin) over the outer layer of the composite and a top layer of the gel coat (white) to ensure longer life in continuous sea water immersion for at least 4 years. However, the study was restricted to only blister formation, but did not indicate change in mechanical properties on sea water aging.

Mechanical properties for short period were investigated by Espinel et al. [28], which revealed that the tensile strength for USP-glass FRP reduced by 20% after 125 days immersion, and interestingly, while the tensile strength attained a constant value after 30 days of saturation, the transverse strength in flexure continued to decrease till 125 days, indicating that the fiber-polymer delamination is more observed if flexural properties are considered. The reason for difference in behavior in these two modes is that the interface delamination affects the bending load-bearing capacity, while in tension, the maximum load is taken by the fiber as such. Unless the fibers are damaged to very high extent as to break down below a critical length, the longitudinal strength will not decrease significantly. This is perhaps the reason for most researchers to measure flexural properties of composites rather than tensile for sea water aging study.

Kootsookos et al. [39] studied the sea water durability of GFRP and CFRP based on USP containing about 32–35% fiber by volume. The flexural modulus of GFRP was about 50% of that for CFRP. The water uptake trend was similar to many other observations, a peak water uptake of about 0.75% for GFRP and 0.5% for CFRP after 16–20 days. However, the water ingress curve had a negative slope after 20 days for both composites. The corresponding flexural modulus of GFRP showed an initial increase, and then finally after 145 days, there was no significant change. Whereas the modulus for CFRP initially decreased and ultimately the reduction is also negligible. It was opined that the weight reduction after peak water uptake is due to hydrolysis and loss of small molecules, but the modulus did not change much for the period of study (145 days). However, the flexural strength of GFRP was seen to reduce considerably, by about 33% but that for CFRP did not change significantly. The performance of the CFRP in sea water aging was seen to be much superior to the GFRP based on polyester resin.

Loos et al. [40] studied hydrothermal effect on USP-based GFRP using distilled water and saturated NaCl solution at 32°C and 50°C. The authors observed that at 32°C, the weight increase is continued with time till saturation value of 3.5–3.6% on 100 days and remained constant thereafter (till 150 days) when immersed in distilled water. For immersion at 50°C, the weight change was having a negative slope after about 50 days. Similar observations were also made by Fraga et al. [41], who studied hydrothermal aging and its effect on interlaminar shear strength and dynamic mechanical properties of GFRP made from isophthalic-acid-based USP with styrene as cross-linker. After 12 days exposure in water at 80°C, the weight change showed a negative slope indicating release of silane coupling agent (sizing of fiber) and also small organic molecules due to hydrolysis of the resin at the elevated temperature. The shear modulus was reduced by about 50% for the composite at 80°C at the end of the study period (1000 h). The glass transition did not change significantly, and the dynamic modulus increased by about 30% but flexural modulus reduced by about 25% at 80°C after about 400 h but stabilized thereafter till 1000 h of study.

Although an USP made of isophthalic acid and neopentyl glycol meets the requirement as a marine-grade resin, as its water resistance is much better than the other USPs, but on a comparison with epoxy resin and vinyl ester resin, the strength of the USP is quite lower, which necessitates a thicker section of a component, say hull of a boat, and consequently there is a possibility of more defects, enhanced water ingress, and faster damage.

5. Diffusion of water in thermosets and FRP

The design of a marine structure is fully dependent on the mechanical properties of the candidate material in various modes. In addition, it must consider the environment in which the object has to perform. Therefore, there is a third consideration of timescale of the service. A very simple example is a static beam under a constant bending load in a building that should carry the load for a long period, for instance, 2–4 decades. Therefore, the design input must be the properties of the material after aging for that service period in the atmospheric environment, especially moisture, carbon dioxide, ultraviolet ray, oxygen, and ozone. While it is not possible to have a data for such a long period for designing an object, it is best to make a prediction of the extent of degradation/aging and degraded properties after a target period of service. This simulation is quite difficult because all environmental and load conditions cannot be simultaneously considered in the mathematical predictive equations. However, a preliminary knowledge or previous study might help in deciding the conditions of fastest degradation due to aging effect. For example, it is known that polypropylene degrades in sunlight due to UV much faster than any other environmental conditions. Therefore, the service life is better decided upon aging under UV of varying intensity.

For marine structures and vessels, the most important considerations to decide the service life are sea water aging, fatigue due to vibration, constant load, and also degradation due to microbial activities. Cyclic sorption-desorption along with a pre-stress was studied by Burla [42], which gave more information on the repeated sorption phenomenon of the cloisite 10A nanocomposites of epoxy, vinyl ester, and unsaturated polyester.

Atmospheric aging due to ozone, UV, etc., is also important for the objects or part of the structures above the water line. In all the factors, sea water aging is most severe because of dissolved salts and alkalinity. The pH of sea water is about 8.3 on an average, and it also contains chlorides, bromides, iodides, sulfates, and carbonates of sodium, magnesium, potassium, calcium, and also traces of heavy metals such as Iron, manganese, cadmium, lead etc. Therefore, diffusion of sea water, and the effect thereof, is the most relevant study for deciding the degradation in mechanical properties of FRP for marine application. It is well known that the extent of sea water uptake and its effect is quite different from potable water or industrial process water.

5.1 Fickian model: constant diffusivity

The diffusion phenomenon in pure thermosets and corresponding FRPs can be generally described by a fundamental theory of diffusion by Fick's Law:

$$\frac{\partial C}{\partial t} = D\left(\frac{\partial^2 C}{\partial x^2} + \frac{\partial^2 C}{\partial y^2} + \frac{\partial^2 C}{\partial z^2}\right) \tag{1}$$

where C is the instanteneous concentration of the diffusing molecule, D is the coefficient of diffusion, commonly called the diffusivity, t is the time, and x, y, z are the three Cartesian coordinates.

Considering only unidirectional diffusion of sea water in a thin panel, (thickness less than 2% of length and breadth), Eq. (1) can be solved to obtain a fractional mass gain (G) at any instant "t" [43]:

$$G = \frac{M_t}{M_\infty} = 1 - \sum_{j=0}^{\infty} \frac{8}{(2j+1)^2\pi^2} \exp\left[-\frac{D(2j+1)^2\pi^2 t}{h^2}\right] \tag{2}$$

where M is the mass, suffixes indicate at any time "t" and the final value, D is diffusivity, j is a simulation factor, h is the thickness. The parameter j actually refines the output more for higher value. For example, $j = 10^6$ gives more accurate result than $j = 10^5$.

The diffusivity can be directly calculated from a simple experiment of water uptake by a panel till saturation, using the following equation:

$$D = \frac{\pi}{16}\left(\frac{M_t/M_\infty}{\sqrt{t}/h}\right)^2 \tag{3}$$

If M_t/M_∞ is plotted against $\sqrt{t}$, D is obtained from the slope of the initial linear portion of the water absorption curve.

Eq. (1) can be approximated as [43]:

$$G = 1 - \exp\left[-7.3\left(\frac{D_x t}{s^2}\right)^{0.75}\right] \tag{4}$$

If the sample is exposed to the sea water on both sides, then $s = h$ (thickness of the panel), if one side is insulated by an impermeable coating, then $s = 2h$.

Rearranging Eq. (4) we can get the time required to attain a certain water content due to unidirectional steady-state diffusion in a thermoset and FRP as:

$$t = \frac{s^2}{D_c}\left[-\frac{1}{7.3}\ln(1-G)\right]^{1/0.75} \tag{5}$$

Eq. (5) is used for predicting the time required to attain any level of water uptake for different thicknesses (s) in cases of a structure with varying contour where the thickness varies sectionwise, and it might not be possible to carry out experiments with all such thicknesses. Typical examples are marine boats, eliptical underwater objects, composite valves in pipelines carrying water, etc. Shen and Springer [44] showed a similar calculation with 12.7 mm thick panel subjected to moisture exposure.

The one-dimensional diffusion equation is valid for thin panels, where the diffusion from edges is not significant. In case of pure thermoset plaques, the edge effect is not very important, but FRP composites are anisotropic materials and hence the edges are to be protected. This is ensured in FRPs by applying the thermoset resin coating on all edges of the panel. However, there can be an edge correction too, to be more precise on unidirectional mass transfer, provided the sample is homogeneous in diffusivity in all directions [43]:

$$D_c = D_z\left(1 + \frac{h}{l} + \frac{h}{w}\right)^{-2} \tag{6}$$

where D_z is the measured diffusivity and D_c is the edge corrected value and l, w and h are length, breadth, and thickness of the panel.

5.1.1 Example

Figure 1 shows an example of water diffusion data for an epoxy GFRP composite of dimensions $l = 100$ mm, $w = 100$ mm, and $h = 4$ mm experimentally determined and a theoretical value calculated by Eq. (4). The Diffusivity calculated from the M_t vs $t^{0.5}$ was corrected for edge error using Eq. (6). The Fickian model clearly does not validate the experimental data, hence the diffusion process might have either more than one diffusive process as the water ingress progresses or there can be an effect of molecular rearrangement in the epoxy thermoset due to absorbed water.

In order to determine the time required for water absorption to the extent of 90% of the saturation for a 12 mm thick FRP panel, assuming identical conditions, Eq.(5) is used and the time calculated as:

$$t = 30,944\ h = 4.58 \text{ years}$$

The life prediction can be done on the basis of the minimum strength required by a designer of the FRP item. Suppose a minimum Flexural strength of 175 MPa is required for the designer to design an underwater vessel hull. The service life of an epoxy-GFRP hull of 12 mm thickness is to be predicted.

Taking the same FRP composition, the laboratory flexural strength data at various times of sea water aging was observed at 35°C for 10,000 h, and **Figure 2** shows the combined data of fractional water absorption and flexural strength with time. The flexural strength measured in 3-point bending test of the original, cured GFRP at 20°C was about 238–245 MPa.

From the source data of **Figure 2**, it is known that the fraction of saturation is 95.4% corresponding to the flexural strength of 175 MPa. Therefore, time required for the 12 mm thick panel at 0.954% saturation as calculated using Eq. (5) is:

$$t = 2397 \text{ days } = 80 \text{ months at a sea water temperature of } 35°\text{C}$$

t = 2397 days = 80 months at a sea water temperature of 35°C

The above solution of prediction is obviously approximate, as the theoretical curve is not in exact agreement with the experimental data till 7500 h (300 days). However, the theoretical prediction of the diffusion curve in **Figure 1** shows better agreement at longer period of exposure. In addition, considering the good fit in **Figure 2** for the fractional saturation (M_t/M_∞) vs. log(time), with R^2 = 0.9786, over the data range of 10–416 days of lab experiment, the data pair of Flexural strength of 175 MPa and 0.954 fractional saturation is fairly accurate compared with experimental value of 0.933 fractional saturation for same strength.

In a different approach, the diffusion-related life estimation can be realized if a time-temperature superposition is done from the data of sea water absorption and a functional property such as strength vs. immersion time at different fixed temperatures of the sea water. In an isothermal analysis with one temperature, the slow relaxation of larger segments of a thermoset polymer (T_g > ambient) is not realized in their contribution to the diffusion and other related properties in long-time aging behavior. There is a slow and continuous physical aging of the matrix at any

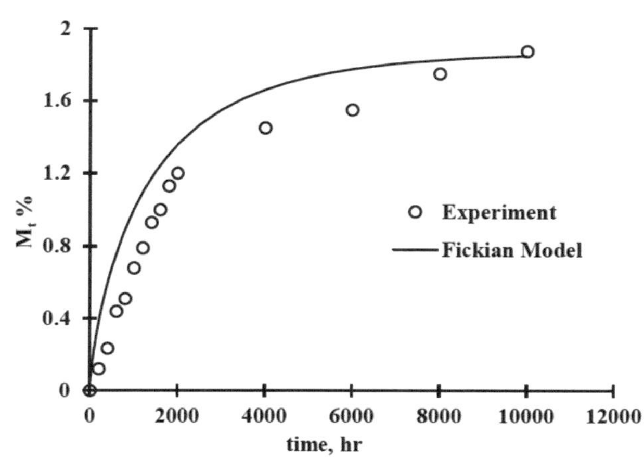

Figure 1.
Typical experimental data and corresponding Fickian model prediction.

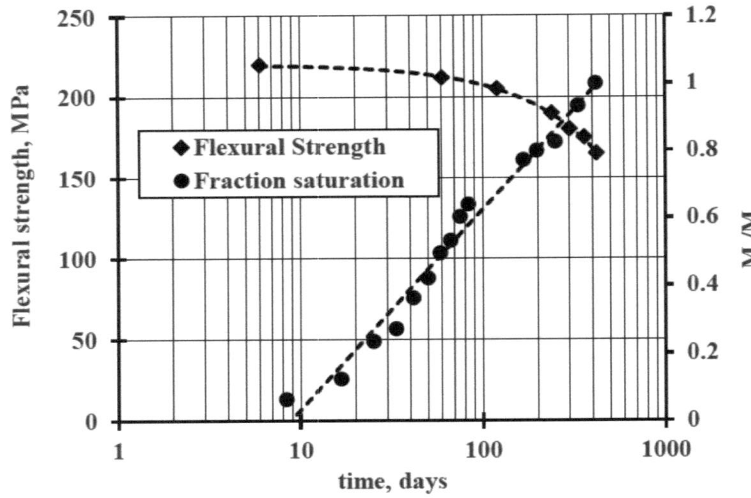

Figure 2.
Flexural strength and fraction of saturation with immersion time in artificial sea water for an epoxy-GFRP at 35°C.

temperature, even below the glass transition. The water absorbed reduces the T_g and if the test temperature is near the modified T_g, the rate of physical aging is enhanced, which may become significant after a long time for a composite of long service life, say more than a decade. It is possible to accommodate such physical process and its effect on properties by resorting to time-temperature superposition principle, since a property at a particular isothermal temperature after a period of aging is same for a different period of aging at another isothermal temperature. This is achieved by shifting the data of an isotherm to another reference isotherm. In classical theory, this relation can be Arrhenius expression as:

$$\tau = \tau_0 e^{E_a/RT} \tag{7}$$

where t is the relaxation time of the polymer segments. Eq. (7) is applicable for high temperatures above glass transitions since the exponential rise in relaxation time is possible when the movement of a segment or a molecule is not hindered by neighboring species. As the polymer is cooled below glass transition, the mobility is hindered by a close approximation of neighboring segments or molecules, and thus the above equation is not valid. In the glass transition region, the relaxation time would change by several decades on even one degree rise in temperature. A rather different expression is Vogel-Fulcher (VF) equation, which is somewhat valid even near glass transition [45]:

$$\tau = A e^{\frac{B}{T - T_0}} \tag{8}$$

However, the value of the constants A, B, and T_0 would change as the temperature is lowered near T_g when more densification of the molecules would take place.

Williams, Landel and Ferry [46] relate the temperature-dependent events such as viscosity, relaxation time, or relaxation frequency with change in fractional free volume of the molecule or segments. The fractional free volume changes linearly with temperature. Accordingly, the relaxation time-temperature relationship is given as the famous WLF equation:

$$\log a_T = \log\left(\frac{\tau}{\tau_g}\right) = \frac{C_1(T - T_g)}{C_2 + (T - T_g)} \tag{9}$$

where $\log(a_T)$ is the shift factor, C_1 and C_2 are constants, T_g is the glass transition temperature, and if it is the reference temperature, then $C_1 = -17.44$ and $C_2 = 51.6$, valid till a test temperature of $T_g + 50°C$, and these values are −8.86 and 101.6 respectively at any other reference temperature up to $T_g + 50°C$, and this is valid till a test temperature up to $T_g + 100°C$.

However, the best process of superposition is to shift the isotherms graphically in a data plot of the property (say strength) vs. time.

A typical *t–T* superposition is shown for a limited time of 8 months as an example to demonstrate the predicted value of a property after aging in sea water. The raw data was plotted in **Figure 3** as a flexural strength vs. time as isotherms at only 20°C, 30°C, 40°C, and 50°C and graphically shifted to the reference temperature 20°C, to predict the value of the strength at longer time than experimental time as an example. For a comprehensive study, longer period and more importantly, higher range of temperature should be used for a prediction at much longer period.

Subsequently, the shifted data are plotted as a master curve with a reference temperature of 20°C as shown in **Figure 4**. The best fit of the shifted data is approximately a second-order polynomial expression here. However, for a long period, the property will vary with the logarithm of time. The shift factors corresponding to the temperatures 30°C, 40°C, and 50°C were used to calculate new time (t_{new}) at the reference temperature of 20°C by following simple equation:

$$\log(t_{new}) = \log(t_{test}) + \log(a_T) \tag{10}$$

The process of determination of shift factor from a graphical shifting is described in Ref. [47]. For example, shift factor $\log(a_T)$ of 50°C is 0.301, and if we take test time at 8 months, then

$\log(t_{new}) = 0.9031 + 0.301 = 1.2041$, therefore, $t_{new} = 16$ months.

This means that the strength at 50°C after 8 months of sea water aging corresponds to 16 months aging in sea water at 20°C. The shifted values can be approximately described by a polynomial fit:

$$\sigma_f = -0.0238t_{new}^2 - 4.0t_{new} + 222.72 \tag{11}$$

where σ_f is the flexural strength in MPa and t_{new} is in months.

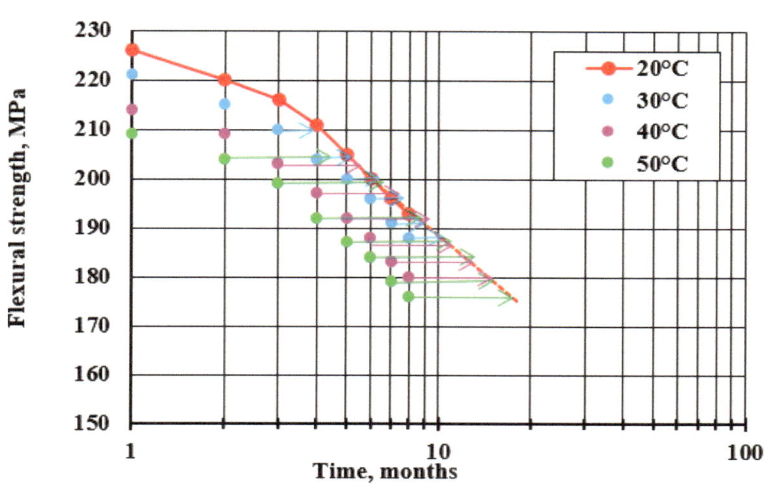

Figure 3.
Isotherms of flexural strength vs. time of immersion of a GFRP based on epoxy resin.

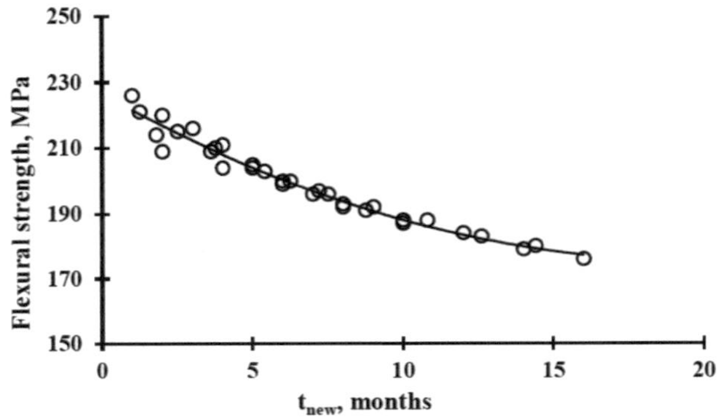

Figure 4.
Master curve for 20°C reference temperature: epoxy-GFRP aging in sea water.

The polynomial fit can be used for determination of the property at extended period too. Therefore, the strength is calculated with Eq. (11) for longer period than the shifted data. **Figure 5** shows the data up to 50 months. The result is obviously an approximation, but gives one the idea of range of the degraded property (strength) for a long exposure time. The validation of the data is not possible unless an experiment is done for the similar period.

Similarly, 10 months data on water uptake by an epoxy-GFRP were studied at limited temperature range of 20°C, 30°C, 40°C, and 50°C. The data were plotted as isotherms and graphically shifted to the reference temperature 20°C. The shift factors were determined, and subsequently new time was obtained using the method already described, and a master curve of water uptake predicted at longer time was obtained. The plot is shown as **Figure 6** here only from 20 months of aging onward till 90 months.

A correlation of these two master curves for prediction of long-term properties as water uptake and flexural strength can be made with some approximation, in this case, because of the limitation of data.

Let us take strength and diffusion data at 48 months from **Figures 5** and **6** respectively. At 48 months, the Flexural strength is 85.6 MPa (calc.) and Water uptake is 6.61% (calc), at a sea water temperature of 20°C.

The example of evaluation of long-term property and water diffusion shown above does not simulate an actual FRP item. In practice, the thicknesses for under-water structures are much higher due to load requirements. Moreover, multiple

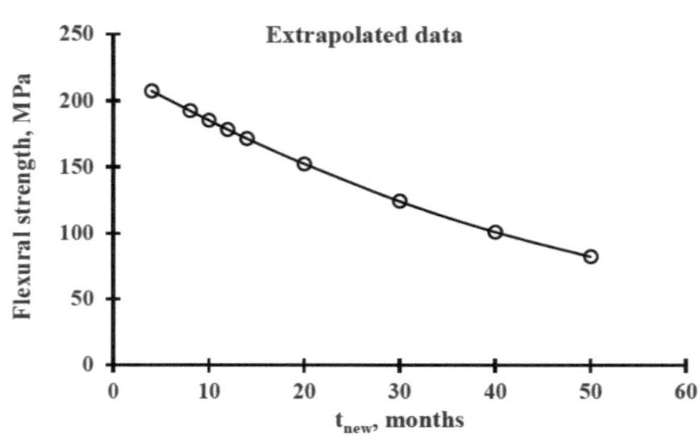

Figure 5.
Extrapolation of the master curve for longer period.

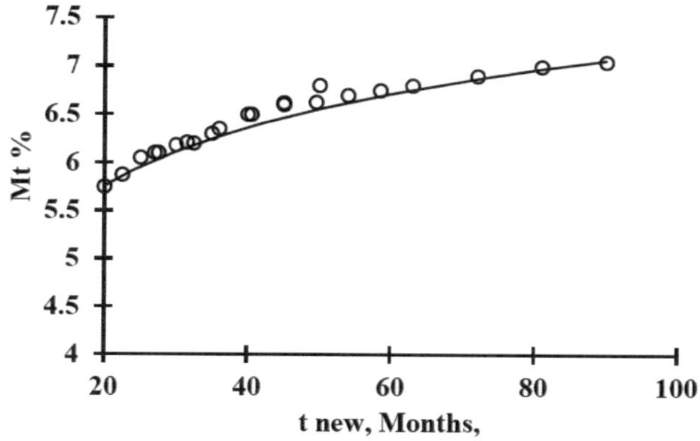

Figure 6.
Long period prediction of water uptake constructed by graphical t-T shift.

types of mats, chopped fibers, fabric with various weaving styles are used in thick composites where FEM analysis is resorted to design the layers.

An approach can be made for life estimation by calculating the diffusion time using Eq. (5) for an FRP of actual size and thickness from the laboratory experiment at different temperature of sea water aging with respect to time. Once a data table is made of M_t% vs. time for the actual size at various isothermal aging temperatures, the data can be used to obtain a graphically constructed master cure following a time-temperature superposition principle for a reference temperature, which is the actual sea water temperature of that geographical region. Since the composites often show dual Fickian behavior or non-Fickian behavior, the data for only long-term study can be taken from the master curve for a good fitting equation. The probability of error is minimized in this process, as graphical shift does not need any assumption such as glass transition temperature, values of activation energy, or the WLF constants, etc. However, a careful experimental determination of the value of diffusivity is required, which is a most critical parameter.

In experiments on diffusion, the panel thickness plays an important part. Although there is an edge correction method available, but it is best to use thin panels of maximum 4.0 mm thickness and edge sealing by a marine-grade vinyl ester resin tissue coat and gel coat of 1.0 mm thickness each. Number of layers of the fabric should be restricted by using fairly thick quality fabric and mats, but not very thick to make the resin infusion difficult. Nevertheless, similar materials such as the resin, curatives, catalysts, and type of mats and fabrics as actual FRP item would be best for a realistic prediction of service life.

5.2 Dual-stage diffusion

After observation of many experimental results on water diffusion process in thermosets and composites, it is certain that the diffusivity is not unique for a case and may vary according to the behavior of the polymer as the process of water ingress progresses. The water diffused in a polymer acts as a plasticizer to change the relaxation process, resulting in swelling, and also initiates some chemical reactions. Karter and Kibler [48] offered a theory that the water absorption is described by a simple diffusion with sources and sinks of diffusing water molecule and that the absorbed water is divided into mobile and strongly bound phases in the polymer. There is a continuous migration from mobile to bound phase and the reverse. There is an equilibrium of this interchange of bound and mobile water. The theory is somewhat similar to Langmuir theory of adsorption-desorption. Considering the probabilities of the interchange of bound and mobile water molecules, the relative mass gain is given by the authors as:

$$\frac{M_t}{M_\infty} = 1 - \frac{\gamma}{\gamma+\beta} e^{-\beta t} - \frac{8\beta}{\pi^2(\gamma+\beta)} \sum_0^\infty (2n+1)^{-2} e^{-k(2n+1)^2 t} \tag{12}$$

where γ is the probability per unit time that a mobile H_20 molecule becomes bound, and β is the probability per unit time that a bound H_20 molecule becomes mobile. The units of both are time^{-1}.

The constant k is given as:

$$k = \frac{\pi^2 D}{h^2} \tag{13}$$

When the exposure time is short, an approximate equation can be used as follows:

$$M_t = \frac{4}{\pi^{3/2}} \left(\frac{\beta}{\gamma+\beta} M\infty \right) \sqrt{kt} \;\; ; \;\; 2\gamma, 2\beta \ll k \;\; ; \;\; t \le 0.7/k \tag{14}$$

Hence, $\beta/(\gamma + \beta)$ can be calculated from the slope of a plot of M_t/M_∞ vs. $t^{0.5}$.

and for a long exposure period, so that kt is much larger than 1. The following approximate equation can be used:

$$M_t = M_\infty \left[1 - \frac{\gamma}{\gamma+\beta} e^{-\beta t} \right] \;\; ; \;\; 2\gamma, 2\beta \ll k \;\; ; \;\; t \gg 1/k \tag{15}$$

Eq. (15) can be rearranged, and after taking logarithm, it becomes:

$$\ln \left(\frac{M_\infty - M_t}{M_\infty} \right) = \ln \left(\frac{\gamma}{\gamma+\beta} \right) - \beta t \tag{16}$$

Eq. (16) represents a straight line with $-\beta$ as the slope of the curve and the intercept would give the value of γ, once β is calculated.

5.2.1 Example

A GFRP based on USP and chopped glass fiber is exposed to artificial sea water at 45°C for 8400 h. The size of the laminate was 80 mm × 12 mm × 4 mm ($l \times w \times h$), and the maximum water uptake was about 4%. The approximate equation Eq. (15) for long-term prediction was used for evaluating the parameters β, γ, and k from the time-dependent absorption data. The values are:

β = 0.0004 h^{-1}, γ = 0.00233 h^{-1} and k = 1.82 × 10^{-05} h^{-1}. The corrected diffusivity (D_c) calculated at the initial slope was 2.95 × 10^{-11} m^2/h.

Figure 7 shows the experimental data and the predicted data of absorption for long-term approximation considering a period beyond 2000 h as the long term.

5.3 Dual Fickian model

Due to the time-varying process of moisture absorption, it is assumed that instead of one constant diffusivity, two diffusivities can be used to describe the long-term water uptake, provided that there is no loss of small molecules as a product of hydrolysis and subsequent leaching out of the experimental panel. The

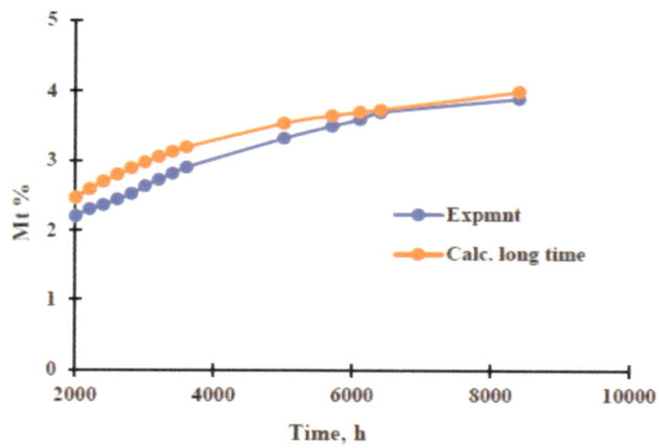

Figure 7.
Long-term water uptake data: experimental and Eq. (15).

initial diffusivity D_1 is determined by the initial slope, and the second diffusivity D_2 is determined by the subsequent part with a distinctly different slope taking only the linear portion. The water uptake is the summation of mass increase for both the diffusivity and the total mass increase M_t at any given time t is given by:

$$M_t = \left(1 - \frac{8}{\pi^2}\sum_{n=0}^{\infty}\frac{1}{(2n+1)^2}e^{\frac{-D_1(2n+1)^2\pi^2 t}{4l^2}}\right)M_{1\infty} + \left(1 - \frac{8}{\pi^2}\sum_{n=0}^{\infty}\frac{1}{(2n+1)^2}e^{\frac{-D_2(2n+1)^2\pi^2 t}{4l^2}}\right)M_{2\infty} \quad (17)$$

where $2\,l$ is the length of the diffusion path (=thickness of the panel), and $M_{1\infty}$ and $M_{2\infty}$ are fractions of final mass uptake M_∞.

A similar expression is a modified Jacob-Jones model [43, 49, 50]:

$$M_t = M_1\left\{1 - \exp\left[-7.3\left(\frac{D_1 t}{b^2}\right)^{0.75}\right]\right\} + M_2\left\{1 - \exp\left[-7.3\left(\frac{D_2 t}{b^2}\right)^{0.75}\right]\right\} \quad (18)$$

Here, $M_\infty = M_1 + M_2$, and b = thickness = $2\,l$ of Eq. (18) above.

Comparing Eqs. (17) and (18), it is more convenient to use the latter, although the equation is an approximate one.

5.3.1 Example

The same water diffusion data of Example 5.1.1, which did not show good fit using the Fickian model with one diffusivity, is tested for the dual Fickian model, taking modified Jacob-Jones expression as in Eq. (18). The diffusivities were calculated as $D_1 = 2.25 \times 10^{-9}$ m^2/h and $D_2 = 7 \times 10^{-10}$ m^2/h. The theoretical and experimental data are plotted in **Figure 8** below. The resulting theoretical prediction is more accurate compared with simple Fickian model (**Figure 1**).

5.4 Diffusion relaxation models

In some models, apart from initial diffusion process of Fickian type, relaxation of the polymer chain segments is also considered, as the water ingress progresses. The water has a plasticizing effect, and hence the relaxational phenomenon, which involves segmental motion of macro-Brownian type, increases with the progress of

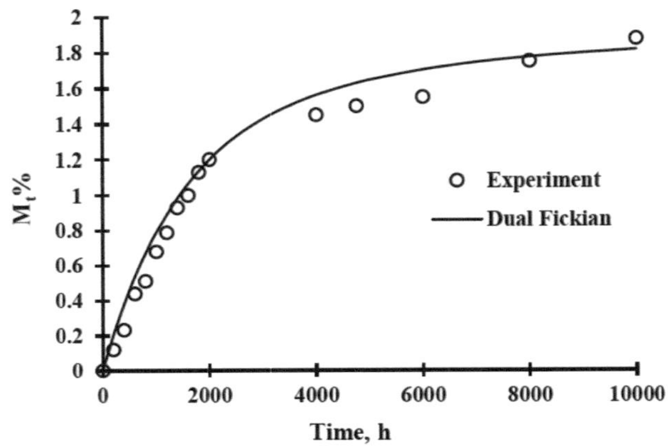

Figure 8.
Dual Fickian model fitted to experimental data of example 6.1.1.

diffusion. The relaxation process in a polymer is related to the slow rearrangements of the chain segments and therefore, distribution of the free volume in the polymer, considering large number of different sizes of the segments in the network. The diffusion and relaxation were combined in a single model by adding the relaxation terms to a classical Fickian diffusion model. The two diffusion processes were assumed to be independent of each other. The mass uptake at any time interval, t, is given by:

$$M_t = M_d(t) + M_{R\infty}(1 - e^{-rt}) \tag{19}$$

where r is a relaxation parameter (inverse of time), M_d is mass uptake for initial phase, calculated by classical equations of diffusion, and $M_{R\infty}$ is the saturation water content due to segmental relaxation process.

The above model is only applicable where the relaxation process is approximately commensurate with the experimental timescale, since a short-term experiment may not result in actual effect of segmental motion and relaxation of a thermoset, which has a very high relaxation time at the experimental temperature.

6. Nanocomposite

Nanometric-sized materials are presently used as reinforcing fillers with polymers. The nanoparticles are defined as those that has at least one dimension below 100 nm. Due to the tiny size, the nanoparticles have very high surface area compared with volume, and hence, their force of attraction with a polymer is much higher compared with common fillers. In addition, the shaped nanoparticles such as rods, platelets, stacked layers, fibers, etc., impart good resistance to diffusion of gas and liquids in polymers.

Needless to say that the intermolecular forces between the nanoparticle and the polymer much depend on homogeneity and polarity. The force of attraction may be Van der Waals, hydrogen bond, polar attraction, dipole-dipole, etc. Some fillers also form covalent bonds too. The secondary valence bonds are physical bonds and are reversible, unlike the covalent bond, which is a chemical bond. Most common nanoparticles are carbon nanotubes, nanorods, nanofibers, graphene and graphene oxide, clays such as montmorillonite, layered silica, nano particles of minerals such as nano titanium dioxide, nano ceramics, etc.

6.1 Common nano fillers

Nano carbons are chemically modified, for example, —COOH functionalized to improve physical bonding with polymers. The reinforcing effect of single-wall carbon nanotube (SWCNT) is much higher than multiwall tubes (MWCNT) because of higher specific surface area.

Graphene and graphene oxides are a new class of plate-type reinforcing nanoparticles, having layer of single graphitic plates, can be physically bunched as 3–8 layers. Graphene is an allotrope of carbon whose structure is a single planar sheet of sp2 bonded carbon atoms that are densely packed in a honeycomb crystal lattice. The graphene can have a spacing of 0.3–0.5 nm between two platelets. Graphene is purer form of carbon, having no organic impurities or functional groups attached, hence their bond with polymers is less intensive than other allotropes of carbon nanoparticles. However, synthesis of graphene from graphite/carbon leads to graphene oxide, which is more polar and can have better bonding with polar resins, which are used for FRP.

Clays are layered silicates, with complex crystal structures. There are different naturally occurring clays such as bentonite, which is kind of rock, mixture of different minerals, including smectite (2:1 layered clay), montmorillonite (dioctahedral), hectorite, (tri-octahedral). The clays are organically modified, for example, ion exchanged with quaternary alkyl ammonium salts, to allow an oligomer molecule to enter in the clay gallery, thus intercalating or even exfoliating the clay. Cloisite is a class of montmorillonite clay, commercially available in various hydrophobicities, and is in the order: Cloisite 15A > 20A > 25A > 10A > 93A > 30B > Cloisite Na^{+}. Typical clay gallery spacing in montmorillonite (Cloisite) is about 1–1.9 nm depending on their structure, and the spacing is increased upon modification and is further exfoliated or intercalated when a low-molecular-weight polymer enters the clay gallery. Typical surface area of a montmorillonite nanoclay of 75–150 nm transverse size is approximately 750 m^2/g.

6.2 Processing of polymer nanocomposites

However, for processing the nanoparticles with an epoxy/unsaturated polyester/vinyl ester oligomer is not easy because of high agglomeration of the nano particles, causing inhomogeneity. In fact, carbon nanotubes cannot be homogeneously mixed with epoxy oligomer beyond 1% without adding any solvent.

Common processing methods are:

1. Ultrasonication: it is effective in low-viscosity fluids. Generally, 40–50 kHz ultrasound is used in a bath containing the polymer mixed with a solvent. It improves dispersion of the nanoparticle by decreasing aggregates or even separates the nanoparticle. For example, SWCNT is mixed with epoxy using dichloromethane as a solvent.

2. Introducing surfactant: composites containing as little as 1 wt% surfactant-dispersed MWCNTs have better homogeneity, resulting in improved interaction between nanoparticle and matrix.

3. Chemically functionalizing the nanoparticle: nanoclays or MWCNTs are organically modified/functionalized leading to an improved dispersion in thermoset forming oligomers such as epoxy, unsaturated polyester, or vinyl ester.

Figure 9 shows a general process flow of polymer-nanocomposite preparation, using ultrasonication. As an example, an epoxy resin with clay is shown here with appropriate processing parameters.

For FRP nanocomposites, it is better to use a pre-bound process rather than mixing the nano material in the resin—which may cause processing difficulties. In this process, the nanoparticle is dispersed in a solvent and sprayed onto the dry fiber mat laid on a steel mesh fitted with a vessel. A vacuum at the bottom of the vessel extracts the gaseous and fluid part and facilitates drying of the mat. Typical parameters for such process for thermoset-MWCNT-based FRP are:

- MWCNT: 1% in acetone
- Air pressure for spray: 0.3–0.4 MPa
- Vacuum: 0.25 atm (absolute)
- Drying after spray: 12 h at 120°C

Figure 10 shows a schematic diagram for the pre-bound process of FRP-nanocomposites.

6.3 Water diffusion in nanocomposites

The barrier property is a function of cross-link density of the thermoset. The free volume for such densely cross-linked network is small, and the chain segments are very stiff. These two factors resist the penetrant transport in the matrix. Since the glass transition temperature is far higher than ambient, the thermoset can be described as an amorphous material trapped as a frozen mass below its glass transition. The dependence of diffusion on cross-linking is reflected in nonlinear behavior and also wide range of diffusivities reported in literature for rigid thermosets and FRPs based on these.

Thermoset nanocomposites were studied by many researchers to observe the effect of inclusion of nanoparticles in highly cross-linked networks, such as epoxy, USP and VE.

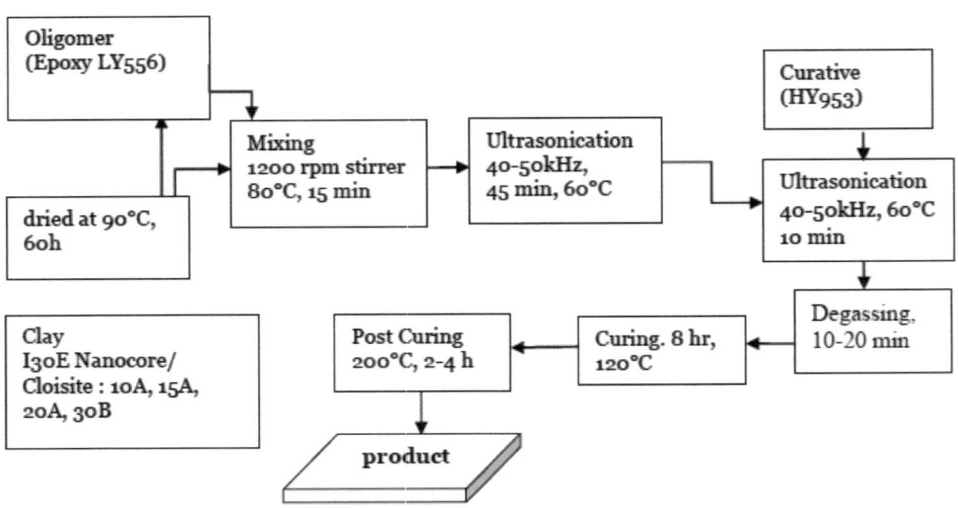

Figure 9.
A typical process for polymer-nanocomposite using ultrasonication.

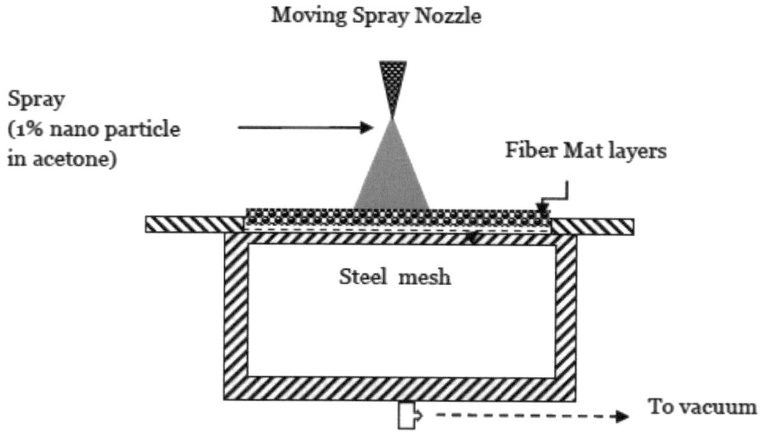

Figure 10.
A typical arrangement for pre-bound process.

Epoxy-amino functionalized carbon nano fiber (CNF) composite was studied by Prolongo et al. [51] and concluded that the CNF effectively controlled the extent of unbound free water, which fills the nano-voids, without swelling (type-I water). However, the final water uptake was not much different than FRP without CNF. Balgis et al. [52] studied MWCNT and milled carbon (spherical graphite and chopped micron scaled carbon fiber) in epoxy resin and found that the dynamic modulus improved by addition of these reinforcements by about 8%. The uptake of water was slow, and the ultimate water was reduced by 12% compared with the neat epoxy resin, and after hydrothermal aging, the dynamic modulus was marginally changed from 2900 MPa to 2700 MPa, but the glass transition temperature changed by about 10–11°C.

Maheshwari et al. [53] studied the effect of nano silica on sea water diffusion of unsaturated polyester resin nanocomposites at varying temperatures (40–60°C) and salinity (0–25%). The effect of inclusion of 3% nano silica in distilled water reduced the saturation water uptake from 0.65 to 0.52% approximately, at ambient temperature, while for a 4% saline water, which is slightly more saline than sea water, the saturation with 3% nano silica is about 0.46%, compared with neat USP at 0.6%. Their study also showed a gradual decrease in saturation of water in nano silica content. See et al. [54] used a organically modified montmorillonite (OMMT) clay treated with a modification agent known as X-treatment using an organically reactive dispersion agent (commercially restricted) in unsaturated polyester resin to make a gel coat with improved barrier property against water immersion/moisture diffusion. It is seen that the moisture uptake increased upon addition of 1% OMMT from 1.74% for without clay to about 2.17%, and with a X-treated OMMT, the figure is about 1.9%. The study clearly shows that the inclusion of the OMMT nanoclay did not reduce the ultimate water uptake, nor it could improve upon glass transition compared with the neat resin after moisture saturation, but the diffusivity was reduced by about 25–30%, and the saturation time is same as the neat resin coating. The life extension by formation of the clay-nano composite cannot be expected to be significant. Shah [55] used two different types of surface-treated OMMT with vinyl resin and studied water diffusion and its effect on properties of the resin-clay nanocomposite. The study indicated similar result as reduction of diffusivity but not the ultimate water uptake and no significant difference in glass transition.

Burla [42] studied the absorption-desorption cycles of water in Cloisite 10A nanocomposites of epoxy, polyester, and vinyl ester thermosets at various values of relative humidity and immersion in water at 25°C under a tensile stress to the extent of 17% of their ultimate tensile strength. Although the diffusivities were reduced upon in addition of the clay, but the ultimate extent of water uptake did not reduce and the time to reach saturation was not improved.

The nanocomposites in all above cases were seen to be non-Fickian in diffusion and in most cases had slightly higher moisture uptake at saturation. This is possibly due to the bound water molecule at the surface of the clay, which is more hydrophilic due to more —OH groups present in the clay compared with that in the resin. A study on fractional free volume and nano hole size distribution was done by Patil et al. [56] with epoxy-Cloisite 10A clay nanocomposites. The authors used Positron Annihilation Lifetime Spectroscopy (PALS) to determine the subnanoscopic free volume in the nanocomposite. The fractional free volume decreased with clay incorporation, but at higher loading, the decrease did not follow a simple linear mixing rule with respect to the volume fraction of the clay, but the reduction was more. This is possibly due to more interaction of the clay with the resin. PALS results showed strong (repulsive) interactions between the clay and the epoxy matrix at lower clay concentrations, which decrease at higher clay concentrations due to the clay-intercalated structure. However, the nanohole size distribution showed an interesting feature. The nanoholes became smaller in size, but the size distribution broadened with respect to nanohole volume beyond the original maximum nanohole volume. There was a net increase in total void volume, although average nanohole volume reduced from 0.075 nm^3 to about 0.05 nm^3 on incorporation of 7.5% cloisite 10A. The reduction of size and increase in overall hole volume are, of course, a function of the clay-resin interaction, for example, in this case it was repulsive.

A study was done by Rath et al. [57] with USP-Cloisite 15A nanocomposites on the reason for reduction of mechanical properties with increase in clay loading. PALS technique was used to find the free volume change on incorporation of the clay. It was seen that clay loading caused an increase in fractional free volume, suggesting a lower chain packing efficiency in these intercalated USP/clay nanocomposites. This could be a reason for higher or at the most similar water uptake on saturation by the clay nanocomposites.

7. Effect of microbial activities on FRP properties

Microorganisms in sea water can settle on structures, such as on metals, FRPs, and on almost all materials. The microbes form a very thin layer of viscoelastic nature as micro-fouling, quite adherent, and this layer is commonly termed as a slime. The slime formation on a substrate can take place within few days of immersion, as there are enormous amounts and varieties of microbes in the sea.

The slime facilitates the settlement of macro-organisms, which is termed as macro-fouling, and most common fouling macro-organisms are bryozoans, barnacles, mollusks, polychaete and other tube worms, zebra mussels, etc. The size of macro foulants is quite large, could be few centimeters even. The slime formation and macro-fouling are highly undesirable for marine vessels and structures, because of many reasons such as evolution of corrosive gases such as hydrogen sulfide, hydrogen, etc., due to metabolism of the organisms, surface roughness of vessels due to macro-fouling, thereby increasing the drag on movement substantially. FRPs are equally vulnerable to such settlements and degradation due to microbial settlement. There are many microbiological studies on the effect of microbes on various materials immersed in sea water. A brief discussion and most important findings are given here for FRP composites.

Little et al. [58] studied the adhesion of the slime on substrates. Gu et al. [59, 60] reported microbial growth and degradation of glass and carbon fibers upon penetration of fungi into the resin matrix. Organic additives to fibers, such as plasticizers and surfactants, may provide nutrients for microbial growth and ultimate degradation as reported by Upsher [61]. Glass fibers are more vulnerable.

Wagner et al. [62] examined carbon fiber-reinforced epoxy (T-300) and a glass (S-2) and carbon fiber (T-300) vinyl ester exposed to microbial culture for 161 days, to study the possible microbiologically influenced degradation. Composites, resins, and fibers were exposed to various microbes including hydrogen producing and sulfate-reducing bacteria (SRB). All types of bacteria colonized surfaces, preferentially on irregularities such as scratches and fiber disruptions. SRB degraded the organic surfactant on glass fibers. Tensile strength of a CFRP of epoxy was reduced on exposure to SRB. The SRB mixed culture did not degrade neat vinyl ester. Degradation of the organic surfactant on glass fibers due to the microbes was observed. Hydrogen-producing bacteria appear to have disrupted fiber-vinyl ester resin bonding with gas production. The study indicated that it is essential for marine application to screen the FRPs against various microbes of sea water before designing the structure.

8. Conclusion

Application of fiber reinforced composites based on thermosets is increasing as the cost and availability of fuel increase with time. This is simply because a lighter marine vessel has the fuel efficiency much higher than metallic vessels. For static RCC construction too, a simple outer jacket of a composite can protect a concrete pillar of off-shore structure for quite a longer period than an exposed RCC. The undersea pipelines are the other areas of potential use of carbon-vinyl ester composites. Nano-carbons can effectively improve the toughness of such items.

A large number of studies are already done to examine the efficacy of using FRPs for use in marine environment considering the chemistry of the matrix resins and relevant properties. Similarly, various fibers were also investigated by many researchers. Recent advancement is focused at incorporation of nanofillers of different chemistry and forms such as nanotubes, fibers, rods, spherical, platelets, etc.

The composites studied so far are widely varying in the resin-fiber ratios, forming sequence, and processing methods. Therefore, the results of each study cannot be fully generalized, but a broad conclusion on quantitative figures of merit can be made for each polymer-fiber combination.

A very general conclusion on durability in marine environment is that vinyl ester resin with carbon fiber is the best choice for applications in static structures, high strength ship components, and commercial speed boats where durability and weight reduction are important. However, for naval ships and submarines, for use in superstructures, the CFRP composites have a problem of radar reflections similar to metallic structures. Modern-day stealth ships exclusively use GFRP since it is radar transparent material (RTM). CFRP is only used in radar absorbing structures (RAS). Internal areas of the vessel can be made with CFRP for better strength and hence reduced weight. However, GFRP has slightly more damping capability than CFRP, which is stiffer. The SONAR Dome can be made with glass fiber or carbon fiber, also with hybrid fiber system. CFRPs are electrically more conductive than GFRP, and hence it is better to use GFRP as inside layers to avoid electrical problems for securing the transducer arrays.

Most studies on durability of FRPs reported so far are either using moisture, or distilled water or artificial sea water, but very rarely natural sea water has been used. Artificial sea water does not simulate the natural sea water. The variations in types of microbes across the world are so much that a result of durability study in sea water at Mumbai coast in India is not applicable in a coast of the United States. There are no comprehensive reports on effect of microbial activity and effect of sea water constituents both considered together to decide a service life of an FRP.

To determine service life of an FRP in marine water, it is required to use the panels immersed in actual sea water using a raft and periodically observing the change in water uptake, chemical groups, mechanical strength, dynamic mechanical properties, surface restructuring, glass transition, etc. Mathematical models commonly used may not be directly applicable for considering the influence of all unforeseen parameters of the sea, but a functional property such as bending strength/modulus can be monitored with time. The data can be superimposed with the similar value of the property with that from a simultaneous laboratory experiment at different temperature as is normally done. The time-temperature superposition will be better used in such cases with graphical shift method to avoid any assumptions. Although the microbe activities are not mapped in temperature scale in such method, it is fairly accurate since the microbe activity is constant due to approximately constant sea water temperature and salinity. The study must be done for at least two cycles of breeding of microbes. This means that the experiment may be only for at least 12 months. Vinyl-ester-based GFRP and CFRP are therefore required to be studied to observe the effects of sew water chemistry and microbiological activity to decide the service life.

Author details

Bikash Chandra Chakraborty
Formerly Scientist, Naval Materials Research Laboratory, Defence Research and Development Organisation, Ambernath, India

*Address all correspondence to: bikash051954@gmail.com

References

[1] Felice R, Nisticò A, Tucci F, Carlone P. Marine application of fiber reinforced composites: A review. Journal of Marine Science and Engineering. 2020;**8**:26, doi:10.3390/jmse8010026

[2] Mouritz AP, Gellert E, Burchill P, Challis K. Review of advanced composite structures for naval ships and submarines. Composite Structures. 2001;**53**:21-41

[3] Galanis K. Hull construction with composite materials for ships over 100 m in length [M.S. thesis]. Massachusetts Institute of Technology: The Department of Ocean Engineering; 2002

[4] Miller AJ. Practical aspects in the construction of large GRP boats. Trans. I Mar E (C) 97. Conf.2. 1984. pp. 7-15

[5] Patri M, Samui AB, Chakraborty BC, Deb PC. Thermal and flammability characteristics of urethane modified polyisocynurate foams based on chlorine containing polyols. Journal of Polymer Materials. 1995;**12**:129

[6] José-Trujillo E, Rubio-Gonza'lez C, Rodrı'guez-Gonza'lez JA. Seawater ageing effect on the mechanical properties of composites with different fiber and matrix types. Journal of Composite Materials. 2019;**53**(23):3229-3241. DOI: 10.1177/0021998318811514

[7] Thomason JL. Glass fibre sizing: A review. Composites Part A Applied Science and Manufacturing. 2019;**127**: 105619

[8] Assbee KHG, Wyatt R. Water damage in glass fibre/resin composites. Proceedings of the Royal Society A. 1969;**312**:553-564

[9] Augl JM, Berger AE. The Effect of Moisture on Carbon Fiber Reinforced Epoxy Composites. I. Diffusion. NSWC/WOL/TR 76-7. White Oak, Silver Spring, Maryland: Naval Surface Weapons Center; 1976

[10] McKague L Jr, Reynolds JD, Halkias JE. Swelling and glass transition relations in epoxy matrix material in humid environments. Journal of Applied Polymer Science. 1978;**22**:1643-1654

[11] Delsai R, Whiteside JB. Effect of moisture on epoxy resins and composites. In: Vinson JR, editor. Advanced Composite Materials-Environmental Effects. ASTM STP 658. Philadelphia Pa. 19103: ASTM; 1978. p. 2

[12] Whitney JM, Browning CE. Some anomalies associated with moisture diffusion in epoxy matrix composite materials. In: Vinson JR, editor. Advanced Composite Materials-Environmental Effects. ASTM STP 658. Philadelphia Pa. 19103: ASTM; 1978. p. 43

[13] Loos AC, Springer GS. Moisture absorption of graphite-epoxy composites immersed in liquids and in humid air. Journal of Composite Materials. 1979;**13**:31-147

[14] Bohlmann RE, Derby EA. Moisture diffusion in graphite/epoxy laminates: Experimental and predicted. In: Proceedings of the 18th Structures, Structural Dynamics and Materials Conference and Conference on Aircraft Composites: The Emerging Methodology for Structural Assurance; March 1977. San Diego, California: AIAA Technical Papers. A: (A77-25726 10.39); 1977. pp. 219-226

[15] Shirrell CD. Diffusion of water vapor in graphite/epoxy composites. In: Vinson JR, editor. Advanced Composite Materials-Environmental Effects. ASTM STP 658. Philadelphia Pa. 19103: ASTM; 1978. p. 21

[16] Marom G, Broutman LJ. Moisture in epoxy resin composites. The Journal of

Adhesion. 1981;**12**(2):153-164. DOI: 10.1080/00218468108071196

[17] Bonniau P, Bunsell AR. A comparative study of water absorption theories applied to glass-epoxy composites. Journal of Composite Materials. 1981;**15**:272-293

[18] Singh KS, Singh PN, Rao RMVGK. Hygrothermal effects on chopped fibre/woven fabric reinforced epoxy composites. Part A: Moisture absorption characteristics. Journal of Reinforced Plastics and Composites. 1991;**10**(5):446-456

[19] Adams RD, Singh MM. The effect of immersion in sea water on the dynamic properties of fibre-reinforced flexibilised epoxy composites. Composite Structures. 1995;**31**(2):119-127

[20] Wood C, Bradley WL. Determination of the effect of seawater on the interfacial strength of an interlayer e-glass/graphite/epoxy composite by in situ observation of transverse cracking in an environmental SEM. Composites Science and Technology. 1997;**57**:1033-1043

[21] Ellyin F, Rohrbacher C. Effect of aqueous environment and temperature on glass-fibre epoxy resin composites. Journal of Reinforced Plastics and Composites. 2000;**19**:1405-1427

[22] Rutowska M, Krasowska K, Heimwoska A, Steinka E, Janik H. Degradation of polyurethanes in sea water. Polymer Degradation and Stability. 2002;**76**:233-239

[23] Ray BC. Temperature effect during humid ageing on interfaces of glass and carbon fibers reinforced epoxy composites. Journal of Colloid and Interface Science. 2006;**298**:111-117

[24] Aktas A, Uzun I. Sea water effect in pinned-joint glass fiber composite materials. Composite Structures. 2008;**85**:59-63

[25] Mourad A-HI, Abdel-Magid BM, El-Maaddawy T, Grami ME. Effect of seawater and warm environment on glass/epoxy and glass/polyurethane composites. Applied Composite Materials. 2010;**17**:557-573

[26] Wei B, Hailin C, Son S. Degradation of basalt fibre and glass fibre/epoxy resin composites in seawater. Corrosion Science. 2011;**53**(1):426-431

[27] Chakraverty AP, Mohanty UK, Mishra SC, Biswal BB. Effect of hydrothermal immersion and hygrothermal conditioning on mechanical properties of GRE composite. IOP Conference Series: Materials Science and Engineering. 2017;**178**:012013

[28] Garcia-Espinel JD, Castro-Fresno D, Parbole Gayo P, Ballester-Muñoz F. Effects of sea water environment on glass fiber reinforced plastic materials used for marine civil engineering constructions. Materials and Design. 2015;**66**:46-50. DOI: 10.1016/j.matdes.2014.10.032

[29] Komorek A, Przybylek P, Kucharczyk W. Effect of sea water and natural ageing on residual strength of epoxy laminates, reinforced with glass and carbon woven fabrics. Advances in Materials Science and Engineering. 2016;**2016**:3754912. DOI: 10.1155/2016/3754912

[30] Suresh M, Babu T, Ravinthiran A, Dhanalakshmi A, Ganapathy S. Effect of sea water environment exposure on glass fiber reinforced polymer composites. IJITEE. 2019;**9**(1):2278-3075

[31] Chakraverty AP, Mohanty UK, Mishra SC, Satapathy A. Sea water ageing of GFRP composites and the dissolved salts. Materials Science and

Engineering. 2015;**75**:012029. DOI: 10.1088/1757-899X/75/1/012029

[32] Murthy HNN, Sreejith M, Krishna M. Seawater durability of epoxy/vinyl ester reinforced with glass/carbon composites. Journal of Reinforced Plastics and Composites. 2010;**29**(10):1491-1499

[33] Menail Y, Mahi AEL, Assarar M. Effect of fatigue testing and aquatic environment on the tensile properties of glass and Kevlar fibers reinforced epoxy composites. Journal of Aeronautics & Aerospace Engineering. 2015;**4**:150. DOI: 10.4172/2168-9792.1000150

[34] Li S, Shuaicheng G, Shi ZJC, Zhu D. The effects of aging in seawater and SWSSC and strain rate on the tensile performance of GFRP/BFRP composites: A critical review. Construction and Building Materials. 2021;**282**(3):122534. DOI: 10.1016/j.conbuildmat.2021.122534

[35] Wonderly C, Grenestedt J, Fernlund G, Cěpus E. Comparison of mechanical properties of glass fiber/vinyl ester and carbon fiber/vinyl ester composites. Composites: Part B. 2005;**36**:417-426

[36] Mungamurugu PK, Marru P, Sardar HH, Neogi S. Long term performance study of glass reinforced composites with different resins under marine environment. Fibers and Polymers. 2017;**18**(1):122-130

[37] Patri M, Chavan JG, Deb PC. A process for preparation of styreneless polyester. Indian Patent No. 290947. Assignee: DRDO, India

[38] Norwood LS. Blister formation in Glass-fibre-reinforced Plastic: Prevention rather than cure. In: Polymers in Marine Environment. Trans. I Mar E (C) 97 1985. Conference No. 2. pp. 23-35

[39] Kootsookos A, Mouritz AP, St John NA. Seawater durability of glass- and carbon-polymer composites. Composites Science and Technology. 2004;**64**:1503-1511

[40] Loos AC, Springer GS, Sanders BA, Tung RW. Moisture absorption of polyester-E glass composites. Journal of Composite Materials. 1980;**14**:142-154

[41] Fraga AN, Alvarez VA, Vazquez A, Osa ODL. Relationship between dynamic mechanical properties and water absorption of unsaturated polyester and vinyl ester glass fiber composites. Journal of Composite Materials. 2003;**37**(17):1533-1574

[42] Burla S. Barrier properties of polymer nanocomposites during cyclic sorption-desorption and stress-coupled sorption experiments [thesis]. West Virginia University: Department of Engineering and Mineral Resources; 1696. Available from: https://researchrepository.wvu.edu/etd/1696; 2006

[43] Maggana C, Pissis P. Water sorption and diffusion studies in an epoxy resin system. Journal of Polymer Science Part B: Polymer Physics. 1999;**37**:1165-1182

[44] Shen C-H, Springer GS. Moisture absorption and desorption of composite materials. Journal of Composite Materials. 1976;**10**:2-20

[45] Roland CM. Characteristic relaxation times and their invariance to thermodynamic conditions. Soft Matter. 2008;**4**:2316-2322

[46] Ferry JD. Viscoelastic Properties of Polymers. 3rd ed. New York: John Wiley; 1980. p. 292

[47] Chakraborty BC, Ratna D. Polymers for Vibration Damping Applications. Amsterdam: Elsevier; 2020. Ch. 3: pp. 243-244

[48] Carter HG, Kibler KG. Langmuir-type model for anomalous moisture diffusion in composite resins. Journal of Composite Materials. 1978;**12**:118-131

[49] Jacobs PM, Jones FR. Diffusion of moisture into two-phase polymers. Journal of Materials Science. 1989;**24**(7): 2331-2336

[50] Jacobs PM, Jones FR. Diffusion of moisture into two-phase polymers. Journal of Materials Science. 1989;**24**(7): 2343-2347

[51] Prolongo SG, Campo M, Gude MR, Chaos-Morán R, Ureña A. Thermophysical characterisation of epoxy resin reinforced by amino-functionalized carbon nanofibers. Composites Science and Technology. 2009;**69**:349-357

[52] Balgis D, Bekas D, Tsirka K, Parlamas A, Ntaflos A, Zafeiropoulos N, et al. Multi-scaled carbon epoxy composites underwater immersion: A durability study. Composite Science and Technology. 2020;**199**:108373. DOI: 10.1016/j.compscitech.2020.108373

[53] Maheshwari N, Neogi S, Praveen Kumar M, Niyogi D. Study of the effect of silica nanofillers on the sea-water diffusion barrier property of unsaturated polyester composites. Journal of Reinforced Plastics and Composites. 2013;**32**(13):998-1002

[54] See SC, Zhang ZY, Richardson MOW. A study of water absorption characteristics of a novel nano-gelcoat for marine application. Progress in Organic Coating. 2009;**65**: 169-174

[55] Shah AP, Gupta RK, Rao GHVS, Powell CE. Moisture diffusion through vinyl ester nanocomposites made with montmorillonite clay. Polymer Engineering & Science. 2002;**42**: 1852-1863

[56] Patil PN, Sudarshan K, Sharma SK, Maheshwari P, Rath SK, Patri M, et al. Investigation of nanoscopic free volume and interfacial interaction in an epoxy resin/modified clay nanocomposite using positron annihilation spectroscopy. ChemPhysChem. 2012;**13**: 3916-3922

[57] Rath SK, Sudarshan K, Patri M, Pujari PK. Accounting for the lack of nano-effect in a thermoset/clay nanocomposite: A positron annihilation study. Journal of Physics: Conference Series. 2015;**618**:012038. DOI: 10.1088/1742-6596/618/1/012038

[58] Little BJ, Wagner P, Maki JS, Walch M, Mitchell R. Factors influencing the adhesion of microorganism to surfaces. The Journal of Adhesion. 2015;**20**:187-210

[59] Gu JD, Ford T, Thorp K, Mitchell R. Microbial growth on fiber reinforced composite materials. International Biodeterioration and Biodegradation. 1996;**37**(3–4):197-204

[60] Gu JD, Lu C, Thorp K, Crasto A, Mitchell R. Fiber-reinforced polymeric composites are susceptible to microbial degradation. Journal of Industrial Microbiology & Biotechnology. 1997; **18**(6):364-369. DOI: 10.1038/sj.jim.2900401

[61] Upsher FJ. Microbial attack on materials. Proceedings: The Royal Australian Chemical Institute. 1976; **43-44**:173-176

[62] Wagner PA, Little BJ, Hart KR, Ray RI. Biodegradation of composite materials. International Biodeterioration & Biodegradation. 1996;**38**(2):125-132

Chapter 10

Mechanical Properties and Chemical Stability of Bathroom Wall Composites Manufactured from Recycle Polyethylene Terephthalate (PET) Mixed with Cocoa Hull Powder

Paul Nestor Djomou Djonga, Ahmat Tom, Hambate Gomdje Valery and Georges Elambo Nkeng

Abstract

The recovery of plastic waste and agricultural residues has led us to develop composites based on polyethylene terephthalate (PET) filled with cocoa shell powder. These shells have been previously treated with the organosolv process to improve the fiber-matrix interaction. The objective of this work is to develop wall covering materials to replace tiles which require a lot of energy and from PET. The composites were made by the method of melt mixing followed by compression molding. The mechanical, physico-chemical properties and stability to environmental conditions were evaluated. The results showed that the incorporation of 20–30% of powder in the matrix made of PET gave rise to a composite material with good properties for application in construction, as a wall covering replacing the tile. The study showed that the optimum powder weight ratio for optimum composite properties was achieved at a powder weight ratio of 30%. The maximum tensile strength of 60.3 MPa, bending strength of 19.5 MPa, impact strength of 10.3 MPa and water absorption of 1.34% were obtained. Compared with ceramic tile, this water absorption test value is within the range and shows that this composite tile is suitable for use as a bathroom tile.

Keywords: chemical stability, cocoa hull powder, composites, mechanical properties, PET

1. Introduction

Plastic consumption and its latter disposal have become a problem due to the high volume of waste and the huge environmental impact they have, not only for the human population, but also for ecological systems [1–4]. Plastic is a versatile material with wide applications. However, it is a material that people do not consume correctly as there are no perceived dimensions on the environmental damage that its use entails

[1, 5, 6]. Per capita plastic consumption continues to rise and remains high in high-income countries, despite obvious contributions to the global issue of plastic pollution [1, 7]. In 2015, the World Bank concluded that if waste generation maintains the same dynamic without adequate actions to improve reuse, unsustainable use, and production, it will have become a health emergency issue in most countries. This is in addition to high greenhouse gas emissions by 2030. Our planet is not capable of digesting the plastic waste generated daily, which will continue to happen. In Cameroun precisely in Yaoundé and Douala town, each person consumes an average of 2 kg of plastic per month, 24 kg per year, which means 1250 Mt. per year in those town [8]. The propagation of plastic waste in the environment constitutes a serious threat to public health because it contains in their structures pollutants and heavy metals that have enormous consequences on the health of living beings [9–11]. With the increase in the volume of waste in our cities, it is more than urgent to consider means of reducing these volumes. As a result, while most of the plastic available today is made from non-biodegradable sources, landfilling using plastic would mean burying the harmful material for a period of time until it naturally degrades. However, their degradation rate and bulky nature create enormous environmental risks. In addition, the mass of plastic waste can hinder the movement of groundwater [12], hence the need to give new life to the waste that would constitute a raw material for the construction industry of building materials. Plastic waste can be reused in several sectors of life. Several processes are used. We will cite thermochemical and thermal processes. A very interesting way to dispose of plastic waste is mechanical recycling, which consists of collection, shredding, and granulation followed by its reintroduction into the manufacture of other plastic products [13]. Given the soaring costs of construction materials, the ambient unemployment that many young people suffer, this valuation sector can create jobs and reduce poverty, which results in the valuation of local raw materials as construction materials. The valuation of plastics in construction materials finds many applications in civil engineering. This on the one hand to the economic advantages is provided by this material, which can act as a binder as a substitute for cement. In addition, several waste management processes are used in civil engineering and have shown that the substitution or use of this material could validly replace binders and would constitute an excellent binder but with low compressive strength values. Much work has been done on the recovery and capacity of plastic waste to be used as binders or as a replacement for cement, see the association of plastic cement in concrete. The most commonly used plastics are polyethylene terephthalate (PET) bottles, polyvinyl chloride (PVC) pipes, high-density polyethylene (HDPE), used plastic waste, expanded polystyrene (EPS) foam, reinforced plastic. Glass (GRP), polycarbonate, recycled thermoplastic polystyrene, polypropylene fiber act as aggregate or mixture in the manufacture of concrete [9].

Reused plastics can be used as reinforcement with natural or synthetic textile fibers in the manufacture of composite materials, the main application of which is the coating of surfaces, walls replacing tiles. Interest in natural fiber-reinforced composites (NFRC) is arising due to their properties of biodegradability, noncarcinogenicity, profitability, and respect for the environment, the absence of health risks, easy collection, and regional availability. They are also used as renewable resource, thus offering a better solution for the sustainability of supply [13]. The different properties of the NFRC show that they would find wide application in the automotive sector, railroad cars, building construction, wall partitions, cabinets, furniture, and packaging manufacturing [6]. NFRCs are viable due to the wide availability of natural fibers and fibers from agricultural residues or mostly ligno-cellulosic can be used as reinforcements [13, 14]. Several authors have embarked on the search for fibers or natural woody materials having properties similar to synthetic fibers from a physico-mechanical point of view that can be naturally or directly used as reinforcement in composites [13, 15].

Indeed, polymers based on synthetic fibers as reinforcements are expensive and have harmful consequences on the environment. Despite the advantages, natural powder has a high water absorption capacity, hydrophilic in nature, which makes it difficult to manufacture composites. The powder swell is softened when in contact with humidity and consequently absorbs water, which contributes to the reduction of their mechanical properties, while its hydrophilic character affects the dispersion of the fibers within the matrix phase. To improve the interaction between the powder and the polymer matrix, natural fibers must be physically or chemically modified to increase their reactivity and their physico-mechanical properties. Much work has been done on composites reinforced with natural fibers such as kenaf, palm oil, bamboo, jute, sisal, coconut, and pineapple fibers. But there is hardly any work in the literature that uses cocoa shell powder as filler. Cameroon is at the forefront of the production and exportation of certain cash crops, such as cocoa (fifth largest producer in the world), bananas, pineapples. It has embarked on a process of modernizing its agriculture and is currently implementing the so-called second generation agriculture, which aims to boost national agricultural production. This generates a lot of agricultural waste. Thus, in 2013, around 700,368 tonnes of cocoa hulls were produced and around 2,900,000 tonnes will be produced by 2022 [16]. Only a small part has been used as fertilizer and animal feed [17], which poses environmental problems, because these cocoa shells therefore pollute the soil and the rivers. This research aims to use the powder from the cocoa shell to strengthen recycled PET. The tensile strength, bending, water absorption, and resistance to acidic and basic conditions will be evaluated in order to measure the physicochemical and mechanical properties of composites.

2. Materials and methods

2.1 Materials

The raw material used in this work consists mainly of waste plastics (PET) used as a binder and cocoa hull powder.

The waste plastic samples were taken from the Center region of Yaoundé, Cameroon. Plastic waste was collected in the Mendong area of Yaoundé. **Table 1** shows us the physical and mechanical characteristics of plastics (PET).

2.2 Methods

2.2.1 Procedure for obtaining cocoa shell powder

After podding, the cockles were cut up and collected in a polythene bag transported to the laboratory, and then dried in a Heraeus brand oven at 45° C until a constant mass is obtained. The dried slices were crushed using a mechanical grinder and sieved with a 1-mm sieve. The powders obtained were packaged in plastic bags and stored at room temperature (25 ± 2°C) until use.

Properties	Absorption rate (%)	Density	Resistance to traction (MPa)	Mass volume (kg/m^3)	Latent heat of fusion (J/g)	Thermal capacity J/kg °C
PET	0.0016	0.93	24	1395	115	1090

Table 1.
Physical, mechanical, and thermal characteristics of PETs.

2.2.2 Physicochemical characterization of cocoa shell powder

2.2.2.1 Water and dry matter content

The water content was determined using the method described by AFNOR (1982) reported by Bachmann et al. [18]. A mass of the fresh sample is dried at 105°C in an oven for 24 h until a constant weight is obtained.

2.2.2.2 Crude fiber content

The determination of the crude fiber content in our powders was determined using performed according to the method described by treating the sample to the boil with sulfuric acid followed by soda. The postharvest by-product obtained is then dried in an oven at 60°C for 24 h, calcined in an oven at 550°C for 4 h with a rate of rise in temperature of 10°C/min and weighed [19].

2.2.2.3 Lignin quantity

The lignin level was determined by the Klason method reported by Monties [20]. The method consists in observing the power of insolubility in an acidic solvent in a concentrated medium and dissolving all the constituents present in the material. The aim of this analysis was to obtain a raw material rich in woody constituent.

2.2.2.4 Reducing sugars quantity

The reducing sugar content was determined using the DNS (acide 2-hydroxy-3, 5-dinitrobenzoïque) method described by Fischer *et al*. [8]. It begins with an extraction of sugars in a hot acidic medium. In a hot alkaline medium, DNS reacts with reducing sugars and changes from its yellow-oxidized form to its orange-reduced form with a maximum absorption at 530 nm.

2.2.2.5 Determination of mass loss

The difference between the mass of the sample before and after the pretreatment represents the mass loss. The sample is dried at105°C before and after pretreatment until a constant mass is obtained.

2.2.2.6 Pretreatment of hulls by the organosolv process

It consists of a batch reactor with a total volume of 100 mL with the useful volume being 80 mL. The experimental protocol was inspired from that described by Nanfack et al. [21]. A mass of 4 g of cocoa shell is mixed with 80 ml of the pretreatment solvent (ethanol-water mixture 52:48 v/v) and loaded into the reactor. The latter is tightly sealed using silicone glue to ensure it is watertight. It is then put in an oven preheated to 200°C. The treatment is stopped by cooling under running water to room temperature. After cooling, the two solid and liquid phases are separated by filtration. The cocoa shell powder thus treated is dried and stored in a glass jar for later use.

2.2.3 Manufacturing method of composite materials

In this study, fine-sized plastic waste is obtained after shredding with a chisel. Plastic waste is then washed with water and dried at room temperature (25°C) for 3 days.

% of CCP	0	10	20	30	40	50
% of PET	100	90	80	70	60	50

Table 2.
Formulation of the different composites according to the addition of the powder.

Composite for PET and cocoa powder: The appropriate with of cocoa powder was added to melt PET in a Haack Rheomix at 225°C. The mixture weighing 200 g was processed for 10 min. The mixture obtained was then hot press at 225°C for 5 min at a pressure of 15 MPa and cooled under pressure. The choice of the temperature of 225° C. was due to the fact that a preliminary sweeping was carried out in order to find the minimum temperature, which makes it possible to develop the composite so as not to alter the powder. From the work of the literature, the melting temperature of PET was around 250° C. Specimens of dimensions 200 × 150 × 8 mm were prepared using a parallelepiped mold. The mold has been polished with a mold release agent to prevent the PET from sticking to it. The process involved melting the grated PET, adding a predetermined proportion of cocoa powder, melt-mixing to form a homogeneous viscous solution, and placing into the prepared mold. Finally, the mold was closed and the samples were cooled to room temperature under a pressure of 15 MPa for 30 min. The specimens of the tiles produced were shaped by sandpaper and used for testing.

Sheets of 3 mm thickness were obtained. The materials were made at different powder concentrations as shown in **Table 2**.

2.3 Mechanical properties and chemical stability of composite

2.3.1 Tensile test

The tensile test evaluation was performed on the Instron 5567 with a climb speed of 50 mm/min. Preparation of samples for testing was performed according to ASTM D638. At the start of the analysis, the specimen lengthens and the resistance of the specimen increases, which was detected using a load cell; the value was recorded until the composite failed. For this, five samples were tested in order to evaluate the standard deviation [22–25].

2.3.2 Flexural test

Mechanical strength is the combination of tensile strength and compressive strength. The tests were carried out on a universal testing machine. The specimens were prepared according to the test standard ASTM D790 of dimension 200 × 30 × 8 mm. These samples of composites produced in varying proportions of PET were tested on a support span of 130 mm according to the standard [13].

2.3.3 Impact test

Charpy impact tests on notched specimens were performed using a model JBS-300 N pendulum impact testing machine. Specimens were prepared according to the ASTM D6110-18 test standard. With a dimension of 50 mm long and a cross section of 24 mm^2, five specimens were tested and one mean value was reported [13].

2.3.4 Water absorption

The water absorption test was performed on parallelepipedal specimens according to ASTM-C373 [22] where the composite samples were weighed and immersed in a beaker containing distilled water and the whole boiled, for 2 h, and allowed to cool for 24 h at room temperature. The specimens were then wiped with a paper towel and weighed.

2.3.5 Chemical stability

The chemical stability will consist in soaking our composite tiles in an acid solution and a base solution with respective pH of 3 and 12 in order to evaluate the loss of mass over a period of 7 days.

3. Results and discussion

3.1 Characterization of cocoa powder hulls

3.1.1 Chemical composition of cocoa powder hulls

As shown in **Table 3** physicochemical properties of cocoa powder hulls is given below.

The dry matter content in our powder was of the order of 88.1%, against 92.1% in organic matter. This high organic matter content shows that cocoa hulls can be an important source of raw material for the production of composites given its organic matter content. The tensile test evaluation was performed on the Instron 5567 with a climb speed of 50 mm/min. Preparation of samples for testing was performed according to ASTM D638. At the start of the analysis, the specimen lengthens and the resistance of the specimen increases, which was detected using a load cell; the value was recorded until the composite failed. For this, five samples were tested in order to evaluate the standard deviation [13].

3.1.2 Thermal analysis

The analyses of the results of the thermal properties of the powder are presented in **Figure 1**. The thermogravimetric analysis (TGA) (in red) provides the loss of mass of the material during thermal degradation. The heat treatment process is that described by Verma et al. [26]. The blue-colored thermogravimetric (DTG) curve gives the shell degradation temperature profile as defined by Balakrishnan et al. [27].

Constituents	Before pretreatment
Dry matter content (g/100* g)	88.1 ± 0.1
Water content (g/100* g)	11.9 ± 0.1
organic matter content (g/100+ g)	92.1 ± 0.2
Ash content (g/100+ g)	7.9 ± 0.2
Reducing sugar content (g/100* g)	0.020 ± 0.003
Crude fiber content (g/100+ g)	40.1 ± 1.2
Lignin content (g/100+ g)	35.3 ± 1.8
Cellulose content (g/100+ g)	41.4 ± 0.3

+—*dry base,* *—*wet base.*

Table 3.
Chemical composition of cocoa powder hulls.

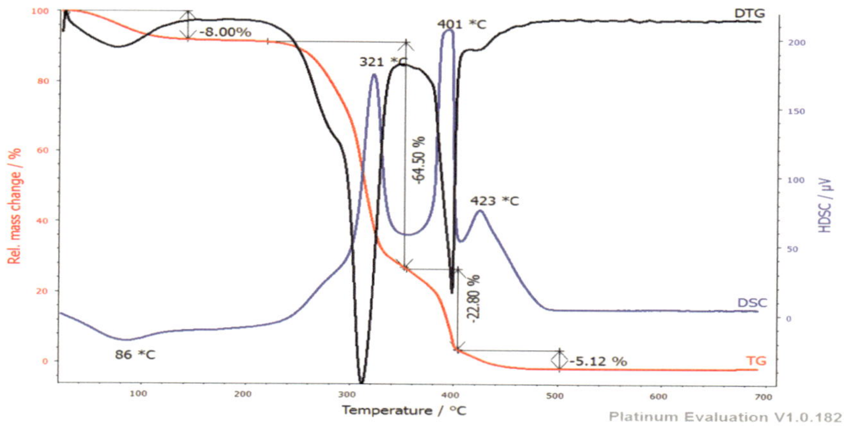

Figure 1.
Thermogravimetric and differential scanning calorimetry of raw cocoa powder hulls.

Careful observation shows that from 440°C all organic matters have been degraded.

The spectrum in **Figure 1** shows four main phases of mass loss. The first mass loss of 8.00% by weight is between 40 and 100°C and has a peak around 86°C; it reflects the loss of free water contained in the cocoa shell powder. The second loss of mass corresponds to a loss of mass of 64.50% observed between 300 and 400°C, with a maximum DTG at 321°C. It is attributed to the destruction of sugars, in particular polysaccharides (hemicelluloses and cellulose) in cocoa shell powder [26]. The third loss of 22.60% observed between 321 and 401°C could reflect the degradation of the cellulose contained in the powder; in fact, cellulose would degrade at around 401°C. Between 401°C and 423°C, a loss of 5.12% is observed with a peak at around 423°C, which would be due to the destruction of lignin at 423°C, these results are consistent with those found by [26].

With regard to the physicochemical properties of these powders, we can conclude that the major compounds are cellulose, hemicellulose, and lignin, which are present in our powder, they degrade with a strong loss of mass [28]. In addition, hemicelluloses decompose at low temperature, due to their weak molecular structure. In addition to their low molecular mass, they have less regular structures in their chains [28]. The curve in **Figure 1** shows four thermal accidents. One endothermic peak is observed at low temperature around 86°C and three exothermic peaks lie between 321°C, 401°C, and 423°C. The peak around 86°C would correspond to the free water contained in the powders, and the other peaks would reflect respectively the degradation of the hemicelluloses, celluloses, and lignin of the cocoa powder hulls [28]. **Table 4** shows the different types of thermal phenomena that occur during the thermal analysis of the sample.

	Temperature (°C)	Partial loss of mass (%)	Total loss of mass (%)	Type of reaction
Powder of cocoa hulls	86	8	100	Endothermic
	321	64.50		Exothermic
	401	22,8		Exothermic
	423	5.12		Exothermic

Table 4.
Types of thermal phenomena.

3.1.3 Infrared Fourier transform analysis

As shown in **Figure 2** and **Table 5**, the IRTF analysis of cocoa hulls powder shows different peak intensities.

Five main groups of adsorption bands emerge, namely those between 3000–3500, 3000–2500, 2000–1500, 1000–500 cm^{-1}. The wavenumber bands between 3000 cm^{-1} and 3500 cm^{-1} is attributed to the vibration of elongation of the linked alcohol OH group. The wavenumber between 2500 cm^{-1} and 3000 cm^{-1} shows the presence of the CH elongation bond. The adsorption band between 2000 cm^{-1} and 1400 cm^{-1} is the characteristic of CO elongation. The absorption band of 1000 cm^{-1} corresponds to the CC bond.

3.2 Mechanical characterization of composite materials

Different formulations were carried out with incorporation rates of the cocoa pod powder from 0 to 50%. The following properties, flexural strength, tensile

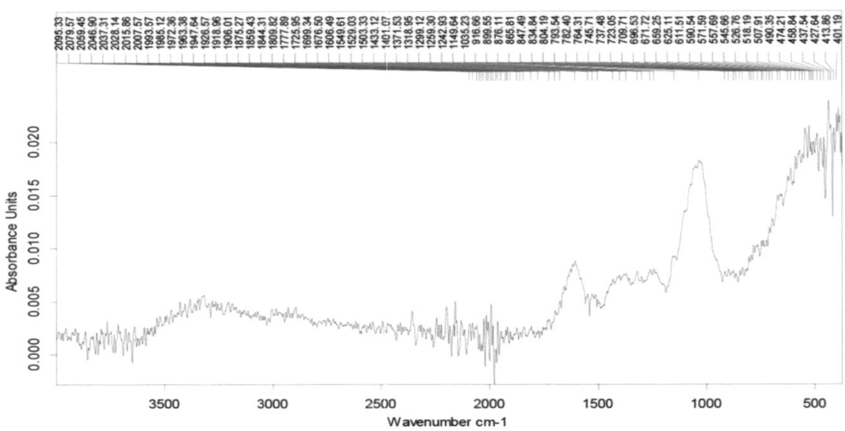

Figure 2.
Infrared of raw cocoa shell powder.

Number of waves (cm^{-1})	Attribution
3315	√ (OH) of lignin and polysaccharides
2927	$\sqrt{}_{as}$ (C–H) of CH and CH_2
1733	√ (C=O) esters and carboxylic acids (xylans, lignins)
1602	Lignin aromatic skeleton
1455	δ_{as} (C–H) from lignin and xylans
1417	δ_{as} (C–H) polysaccharides
1371	δ_{as} (C–H) of cellulose and hemicelluloses
1317	CH_2 from cellulose and hemicellulose
1234	√ (C–O) of methoxy groups of lignin
1025	δ (C–O) of polysaccharides δ (C–OH) of cellulose
896	δ (CH) of cellulose
600	δ (CO–H) of cellulose

√ is the strain in vibration and δ is the strain in rotation.

Table 5.
IR spectrum of cocoa shell powder: the characteristic infrared vibration bands, relating to CCP.

%CCP	Tensile strength (MPa)	Flexural strength (MPa)	Impact strength (MPa)	Water absorption (%)
10	55.8	10.6	8.9	1.18
20	58.2	11.9	7.4	1.20
30	60.3	19.5	10.3	1.34
40	56.1	12.3	7.6	1.49
50	43.9	9.8	6.9	1.51

Table 6.
Mechanical properties of composite.

strength, impact strength, and water absorption, have been used to define the properties of composites. As shown in **Table 6**, the mechanical properties of composites are given.

3.2.1 Tensile strength

The tensile strength was studied as a function of the rate of incorporation of the cocoa powder into the PET. As shown in **Table 6**, that the tensile strength of the composite increases with the amount of load introduced 10–30% and is equivalent to a value of 55.8–60.3 MPa, respectively, but when the level of 30% powder is reached, this value drops from 60.3 to 43.9 MPa. Observation of the table indicates that the greatest value of the tensile strength is reached after adding 30% of powder. This high value of the resistance results in a reinforcement of the PET, which has become more rigid and can withstand large loads. The small size and their particle size distribution in the composite allow good adhesion and contribute to the densification of the composite. Beyond 30% of the powder content, the resistance to decrease that would lead to believe that the PET matrix was not sufficient to cross and homogeneously fluidize the surface of the composite creating pores, which would contribute to the low resistance to traction.

3.2.2 Flexural property

Analysis of the table reveals that the flexural strength increases with the addition of the powder. The incorporation rate of 10 to 30% made it possible to go from a value of 10.5 to 19.50 (MPa). This could be explained by the fact that an increase in the loading rate would lead to the presence of a high rate of loading on a cross section of the composite and therefore increasing the flexural strength. We also note that beyond 30% of the incorporation rate, there is a decrease in the value of the flexural strength. This drop could be explained by poor dispersion of the powder in the matrix and this observation is confirmed by the tensile strength where we see similar results.

3.2.3 Impact strength

Table 6 presents the values of the Charpy impact resistance (CIS) of non-notched samples of PET composites with a proportion of cocoa shell powder varying from 10 to 50%.

For a load rate below 30%, there is an increase in the CIS, which reaches its highest value 10.3 (MPa) at 30% CCP. Beyond 30% the CIS decreases, because the addition of more powder creates regions of stress concentrations that require comparatively less energy to initiate a crack; this reduction could also be explained by the fact of the low resistance of the interface between the powder and the matrix. During impact, most of the energy absorbed is used to increase the distance between the load and the die.

Day	0	2	4	6	8	10
30% CCP at pH 3.2	150	143	141	139	139	138
30% CCP at pH 6.79	150	148	147	147	147	147
30% CCP at pH 12.4	150	151	151	150	150	150

Table 7.
Weight loss of specimen in nitric acid and sodium hydroxide solutions (g).

3.2.4 Water absorption

As shown in the **Table 6** the filler content has a notable effect on water absorption. After 24 h of immersion in water, the lowest water absorption rate was reached with a load proportion of 10%. The incorporation of cocoa shell powder would be help to increase the percentage of water absorption up to a certain content, which means that more hydrogen bonds are formed between the water molecules and the OH group present in the powders. Similar results were observed by Huner et al. who reported that the rate of water absorption increased with increasing fiber content [29, 30]. This could be explained by the formation of a less surface interaction between the matrix and the powder during mixing, resulting in a higher water absorption [31]. In addition, PET would constitute an ideal binder for the compaction of this powder and reduce water absorption. This low water absorption value could improve the physico-mechanical properties.

3.3 Chemical stability

As shown in **Table 7** that after 10 days of impregnation in acidic medium and neutral medium, the composite material undergoes a mass loss of 8 and 2%, respectively, whereas in a basic medium no loss is observed. This result reflects the good chemical stability of the composites under extreme conditions and shows that the material can indeed be applied as a tile coating for the walls of toilets and showers.

4. Conclusion

The chapter studied the chemical stability and mechanical properties of bathroom wall composites manufactured from recycle polyethylene terephthalate mixed with cocoa hull powder. Organosolv treatment of cocoa powder hulls enhanced the quality of the cocoa powder hulls by removal of hemicellulose and lignin thereby increasing the cellulose composition. The Organosolv treatment gave a more thermally stable composite material. The addition of cocoa powder hulls produces an environmentally friendly material. It is clear that cocoa powder hulls have great potential as a filler and reinforcement for composites requiring similar properties to PET. The recycled PET/30% cocoa hull powder blend has higher tensile, flexural, and impact strength compared to other blends. Recycled PET/30% cocoa hull powder blend was also found to have good properties in terms of stability. The manufactured composites showed improved mechanical properties with respect to the powder content increase, showing the reinforcement potential of cocoa powder. It is thus possible to consider a prospective industrial use of this agricultural waste, for instance, for the manufacture of bathroom wall tiles in construction industry in Cameroon.

Conflicts of interest

The authors declare that they have no conflicts of interest.

Data availability

All relevant data to the manuscript have been included.

Author details

Paul Nestor Djomou Djonga[1*], Ahmat Tom[2], Hambate Gomdje Valery[3] and Georges Elambo Nkeng[4,5]

1 Faculty of Science, Department of Chemistry, University of Maroua, Maroua, Cameroon

2 Department of Energy Engineering, University Institute of Technology, University of Ngaoundere, Ngaoundere, Cameroon

3 Departement of Textile and Leather Engineering, National Advanced School of Engineering of Maroua, University of Maroua, Maroua, Cameroon

4 Faculty of Science, Department of Chemistry, University of Buea, Buea, Cameroon

5 Ecole Nationale Supérieure des Travaux Publics, Yaounde, Cameroon

*Address all correspondence to: djomoupaul@gmail.com

References

[1] Sanchez-Echeverri LA, Tovar-Perilla NJ, Suarez-Puentes JG, Bravo-Cervera JE, Rojas-Parra DF. Mechanical and market study for sand/recycled-plastic cobbles in a medium-size Colombian city. Recycling. 2021;**6**:17

[2] Cordier M, Uehara T. How much innovation is needed to protect the ocean from plastic contamination? Science of the Total Environment. 2019;**670**:789-799

[3] Rai PK, Lee J, Brown RJC, Kim KH. Environmental fate, ecotoxicity biomarkers, and potential health effects of micro- and nano-scale plastic contamination. Journal of Hazardous Materials. 2021;**403**:123910

[4] Ribeiro-Brasil DRG, Torres NR, Picanço AB, Sousa DS, Ribeiro VS, Brasil LS, et al. Contamination of stream fish by plastic waste in the Brazilian Amazon. Environmental Pollution. 2020;**266**:115241

[5] Beaumont NJ, Aanesen M, Austen MC, Börger T, Clark JR, Cole M, et al. Global ecological, social and economic impacts of marine plastic. Marine Pollution Bulletin. 2019;**142**:189-195

[6] Rhein S, Schmid M. Consumers' awareness of plastic packaging: More than just environmental concerns. Resources, Conservation and Recycling. 2020;**162**:105063

[7] Barnes SJ. Out of sight, out of mind: Plastic waste exports, psychological distance and consumer plastic purchasing. Global Environmental Change. 2019;**58**:101943

[8] Souop Tagne Alain Andrey. Modeling of biogas production by the particle swarm method: Application to cocoa hulls [thesis presented in view of obtaining the Master's degree in Process Engineering]. p. 99

[9] Kamaruddin MA, Abdullah MMA, Zawawi MH, Zainol MRRA. Potential use of plastic waste as construction materials: Recent progress and future prospect. IOP Conference Series: Materials Science and Engineering. 2017;**267**:012011

[10] Saikia N, de Brito J. Use of plastic waste as aggregate in cement mortar and concrete preparation: A review. Construction and Building Materials. 2012;**34**:385-401

[11] Kamaruddin MA et al. Resource recovery from municipal solid waste by mechanical heat treatment: An opportunity. In: AIP Conference Proceedings. 2017

[12] Silva R, De Brito J, Dhir R. Properties and composition of recycled aggregates from construction and demolition waste suitable for concrete production. Construction and Building Materials. 2014;**65**:201-217

[13] Gebremedhin N, Rotich GK. Manufacturing of bathroom wall tile composites from recycled low-density polyethylene reinforced with pineapple leaf fiber. International Journal of Polymer Science. 2020;**2020**: 2732571. 9 pages

[14] Cyras VP, Iannace S, Kenny JM, Vázquez A. Relationship between processing and properties of biodegradable composites based on PCL/starch matrix and sisal fibers. Polymer Composites. 2001;**22**(1):104-110

[15] Barkoula NM, Garkhail SK, Peijs T. Biodegradable composites based on flax/polyhydroxybutyrate and its copolymer with hydroxyvalerate. Industrial Crops and Products. 2010;**31**(1):34-42

[16] FAO. Bioenergy and Food Security Rapid Assessment (BEFS RA): BRIQUETTES. Rome: FAO; 2014. 43 pages

[17] Adabe KE, Lionelle Ngo-Samnick E. Cocoa Production and Processing. Douala (Cameroon): Pro-Agro Collection; 2014. 44 Pages

[18] Bachmann N, Deront M, Fruteau H, Holliger C, Membrez Y, Wellinger A. Optimization of standardized digestibility tests in batch reactors. Bern: Federal Office of Energy SFOE Energy Research Program, 1015 Lausanne: Laboratory of Environmental Biotechnology (LBE); 2011

[19] Sakala JM, Loumouamou BW, Bopoundza FC, Mampouya D, Mouloungui Z. Bioavailability of Epoxidized Triglycerides in Seed Oil of Alchornea cordifolia (Schumach. &Thonn.) Müll. Arg. from Congo-Brazzaville, Engineering and Technology. 2021;**18**(3):92. DOI: 10.19026/rjaset.18.6068

[20] Monties B. Determination of insoluble lignin in an acid medium: Influence of pre-treatment by acid hydrolysis on Klason lignin from wood and straw Agronomie. EDP Sciences. 1984;**4**(4):387-392

[21] Nanfack NC. Production of biogas from cocoa waste [Master's thesis in process engineering and engineering]. Cameroon: ENSAI, University of Ngaoundéré; 2016. 96 p

[22] ASTM-C373. American Standard and Testing Materials. C373-88 (ASTM). Standard Test Method for Water Absorption, Bulk Density, Apparent, Porosity, and Apparent Specific Gravity of Fired White ware Products. 1999

[23] Daud Z, Kassi AS, Aripin AM, Awang H, Hatta MZ. Chemical composition and morphological of cocoa pod husks and cassava peels for pulp and paper production. Australian Journal of Basic and Applied Sciences. 2013;7(9):406-411

[24] Vriesmann LC, Amboni RD, Petkowicz CL. Cacao pod husks (*Theobroma cacao* L.): Composition and hot-water-soluble pectins. Industrial crops and products. 2011;**2011**(34): 1173-1181

[25] Bai YY, Xiao LP, Shi ZJ, Sun RC. Structural variation of bamboo lignin before and after ethanol organosolv pretreatment. International Journal of Molecular Sciences. 2013;**14**:21394-21413

[26] Verma D, Senal I. Natural fiber-reinforced polymer composites. Biomass, Biopolymer-Based Materials, and Bioenergy. 2007;**44**:129

[27] Balakrishnan P, John MJ, Pothen L, Sreekala MS, Thomas S. Natural fibre and polymer matrix composites and their applications in aerospace engineering. In: Advanced Composite Materials for Aerospace Engineering. Sawston, Cambridge, UK: Woodhead Publishing; 2016. pp. 365-383

[28] Kumar R, Ul Haq MI, Raina A, Anand A. Industrial applications of natural fibre-reinforced polymer composites–challenges and opportunities. International Journal of Sustainable Engineering. 2019;**12**:212-220

[29] Huner U. Effect of water absorption on the mechanical properties of flax fiber reinforced epoxy composites. Advances in Science and Technology Research Journal. 2015;**9**(26):1-6

[30] Zahari WZW, Badri RNRL, Ardyananta H, Kurniawan D, Nor FM. Mechanical properties and water absorption behavior of polypropylene/ijuk fiber composite by using silane treatment. Procedia Manufacturing. 2015;**2**:573-578

[31] Dhakal HN, Zhang ZY, Richardson MOW. Effect of water absorption on the mechanical properties of hemp fibre reinforced unsaturated polyester composites. Composites Science and Technology. 2007; **67**(7-8):1674-1683